Pacific Coast
COMMUTER RAILROADS
From San Diego to Anchorage

Patrick C. Dorin

Iconografix

Iconografix
PO Box 446
Hudson, Wisconsin 54016 USA

Library of Congress Control Number: 2008930631

ISBN-13: 978-1-58388-221-4
ISBN-10: 1-58388-221-9

08 09 10 11 12 13 6 5 4 3 2 1

Printed in China

Cover and book design by Dan Perry

Copyedited by Andy Lindberg

On the cover: A northbound Coaster between San Diego to Oceanside rolling along the Pacific Coast. *Steve Glischinski*

TABLE OF CONTENTS

DEDICATED TO WILLIAM J. DORIN

Brother, University Professor, Author, Researcher and Friend

ACKNOWLEDGMENTS

This book on the new commuter rail services covering the West Coast of the United States and Canada would not have been possible without the kind assistance and wisdom of many people from the commuter rail authorities, passenger car builders and several railroad researchers and photographers.

Following are the people from the Transportation Authorities and Car Builders:

Jim Blasingame and Bruce Carr with the Alaska Railroad, Hubert Hanrahan, Jr. and Stacey Mortensen for the Altamont Commuter Express; Christine Novotny and Stephanie Lawrence for Tri-Met (Portland) and Ann H. Madden, Washington County, Oregon; Jim Moore, Bruce Grey and Mike Bergman for Sounder; Sarah Benson for Coaster; Ed Pederson for Metrolink; Janet McGovern and Cathy Blair for Caltrain; Kyla Daman-Williams for West Coast Express; and Cliff Black for Amtrak; Yasuhiko Mochizuki, Nippon Sharyo, and Arthur Rader for Colorado Railcar Manufacturing.

Erik Frodsham, William S. Kuba, Wilbur C. Whittaker, Anna Whittaker, John Sward, Rosalie Sward, David Sward, Steve Glischinski, Robert Larson, Jeff Koeller and Dan Mackey assisted with both photographs and information regarding the commuter rail services. Should anyone's name been inadvertently missed in the listings above, we trust it will be at the appropriate location within the book.

To each and everyone, thank you for your time, kind assistance and wisdom.

INTRODUCTION

The Pacific Coast Commuter Railroads is about the newest chapter on rail passenger services to be created in North America. The geographical territory for the topic stretches from San Diego to Vancouver, British Columbia, and all the way to Anchorage, Alaska. With the exception of one commuter rail route, the Southern Pacific between San Francisco and San Jose, all of the commuter routes are new and were in the planning stages and developed since the late 1980s/early 1990s and into the first decade of the 21st Century. The purpose of this book is to look at the new systems currently in operation or as they will be between 2008 and 2011.

Since the late 1980s, the new commuter train services began a new level of growth and development. Amtrak has also participated in the commuter rail services in three areas on the west coast, Seattle, Los Angeles and San Diego, with new ticketing arrangements. Commuters also use Amtrak's Capitol Corridor in the Sacramento–San Francisco area.

The new commuter train services not only provide new and more effective and relaxing ways to travel to and from work, but also for shopping, recreation and for many other activities. One example is the schedules for weekend swimming along the Pacific Coast in the Los Angeles–San Diego area. Add in sporting events, concerts, and plays; and there is a wide variety of special services.

Rail passenger service is the only energy-efficient solution to the pollution and congestion to be found in many locations throughout North America. Unfortunately, too many people somehow think railroads are past history, and the only way to solve the problem is to build more highways and parking ramps downtown. Frequently, we hear that the entire congestion problem could be solved by simply adding more lanes to the expressways. However, that in turn causes new levels of problems and accident rates near entries and exits as drivers attempt to change lanes to secure the appropriate route for entry or exit. More concrete, by the way, has another destructive outcome because the additional lanes inhibit the flow of rain water into the ground. More parking ramps destroy even more business locations which depletes the tax base and reduces the number of jobs in the area—which adds more to urban sprawl and the problem keeps getting worse. And this doesn't even take into consideration the rocketing price of gasoline and daily parking fees.

Another concept frequently overlooked is the interdependency between the central cities and the suburban areas. When the cities lose their economic base to highways and parking ramps, the economic health of the entire area suffers.

Fortunately, many of the Pacific Coast metro areas have recognized this situation. This is the primary reason for the new commuter train services. Each of the following chapters describes the individual commuter systems, including the territory served, the extent of train services (such as frequency, time periods, and running times), other transit connections including light-rail and bus systems, motive power and the passenger equipment.

The new services on the Pacific Coast are providing North America with new knowledge levels for solutions for reducing the need for gasoline, congestion, pollution, and road rage. With the reduction of stress for driving to work, commuter rail adds new levels of productivity at the work place and less stress at home. And there are even more benefits because commuter railroad lines as well as light rail create new levels of a sense of community, contribute to the economic growth and development of the area and are environmentally friendly. It is a lesson that needs to spread rapidly across North America. Thus we can say the Pacific Coast Commuter Railroads are a bright star on the horizon.

Patrick C. Dorin
November, 2007

SP train 229 was a Sacramento to Oakland run going back to the 1930s and 1940s. The train provided coach service as well as handling mail and express. Note the three head-end cars behind the 4-6-2 Pacific, No. 2405. This train is on the future route of Amtrak California's Capitol Corridor route between San Jose, Oakland and Sacramento, the subject of Chapter 7. The steam power illustrated here was also part of the motive power used on the commuter trains during the "Steam Era." *William S. Kuba Collection*

Chapter 1
THE SOUTHERN PACIFIC

The Southern Pacific was once the only full railroad commuter train service in California, although there were a number of other routes such as the Pacific Electric. In fact, San Francisco can say it was the only city west of Chicago to have a full service system for the 46.9-mile route between the Bay Area and San Jose.

The Southern Pacific carried about 11,000 passengers per weekday during the 1960s. The company operated 44 weekday trains, which reduced to 24 on Saturdays and 18 on Sundays during this time period.

The rush hour train service consisted of 12 arrivals in downtown San Francisco during the 6:00 to 9:00 a.m. period. There were 10 departures between 4:00 and 7:00 p.m. The frequency of the rush hour train service varied from 3 minutes to 19 minutes between trains. The non-rush hour service frequencies varied from 55 minutes to 2 hours, 35 minutes. The weekday service began at 5:05 a.m. and ran until 12:35 a.m. All of the trains, with the exception of one rush hour run, arrived and departed at San Jose.

During the late 1960s and early '70s, the SP operated and maintained 55 standard coaches and 46 double deck streamlined cars. The double deck cars, or gallery cars as they are known, had a seating capacity of 160 passengers. All of the non rush hour trains were equipped with streamlined gallery coaches. One

SAN JOSE TO SAN FRANCISCO

Southern Pacific Peninsula timetables illustrate the train schedules for weekdays and Saturday and Sunday beginning in January 1968.

SAN FRANCISCO TO SAN JOSE

PENINSULA ZONE FARES

To determine your fare, simply select the zone from which you will commute regularly (remember, you can ride to or from all stations in that zone). Read to right under the type of ticket you prefer to use. This figure is your Zone Fare. You may ask for your ticket by zone and color.

	BETWEEN SAN FRANCISCO AND	5-Day Monthly Commutation ①	Monthly Commutation (Every Day) ②	Weekly Commutation (7-Day) ③	70-Ride "Family" Ticket	Single Ride Tickets One-Way Fares	Single Ride Tickets Round-Trip Fares
RED ZONE 1	So. S. F. San Bruno Millbrae	$17.00	$18.30	$4.80	$12.00	$0.70	$1.30
GREEN ZONE 2	Broadway Burlingame San Mateo Hayward Park	20.50	22.20	5.70	14.40	.90	1.65
ORANGE ZONE 3	Hillsdale Belmont San Carlos Redwood City	24.00	26.10	6.60	16.20	1.10	1.95
BLUE ZONE 4	Atherton Menlo Park Palo Alto California Ave.	27.50	30.00	7.80	18.00	1.35	2.50
YELLOW ZONE 5	Castro Mountain View Sunnyvale	31.00	34.20	9.00	19.80	1.60	2.85
BROWN ZONE 6	Santa Clara College Park San Jose	33.50	36.60	10.20	21.00	1.75	3.10

① Good for unlimited number of rides Monday thru Friday inclusive.
② Good for unlimited number of rides daily.
③ Good for unlimited number of rides Sunday to Saturday inclusive.

Southern Pacific
PENINSULA
Time Tables

SAN FRANCISCO
SO. SAN FRANCISCO
BURLINGAME
SAN MATEO
REDWOOD CITY
PALO ALTO
CALIFORNIA AVENUE
SAN JOSE
LOS GATOS

Subject to change without notice

January 9, 1968

Fairbanks-Morse 2,400-horsepower road switchers with steam generators were typical power on the SP commuter trains. The 3034 is handling train No. 133 en route to San Francisco with a consist of six coaches, all painted in the two-tone grey color scheme, which was typical on the SP during the 1940s and '50s. No. 133 is rolling through South San Francisco on April 22, 1966. *Photo by Friedrich, William S. Kuba Collection*

interesting point for model railroaders is that Walthers has produced the last group of Southern Pacific bi-level gallery coaches (number series 3731 to 3745) in HO gauge.

The motive power operated on the Peninsula Commuter service consisted of GP9s, Fairbanks-Morse 2400 HP road switchers and EMD F7s, E8s and even FP7s. The SP trains were operated in the conventional manner as the company did not adopt the push-pull concept pioneered by the Chicago and North Western and other Chicago commuter railroads.

The Southern Pacific lost money on the commuter passenger services, which was common for virtually all of the railroads in North America. Two excep-

tions were the Chicago and North Western and the Illinois Central services in Chicago during various periods of years in the 1960s. The SP filed a petition with the California State Public Utilities Commission to discontinue the service in 1977. However, the train services were and are crucial to the future development of the entire Peninsula Area, not to mention the alleviation of traffic congestion and pollution. This eventually led to new developments in 1980 with state agencies that led to a new organization, CalTrain, new equipment, track changes and expansions, and an extension of the service further south to Gilroy over the SP route. CalTrain took over the operations on the Southern Pacific, which is covered in Chapter 5 of this book.

Commuter Passenger Equipment—1960s and 1970s

Type	Number Series	Remarks
Gallery Coaches	3700 to 3730	Originally painted in the two-tone grey scheme. 145 seating capacity
Gallery Coaches	3731 to 3745	The last new group of equipment painted in the grey scheme
Heavyweight Coaches	2085 to 2159	96 seating capacity

Train 138 is en route to San Jose as it blasts out of the tunnel at Bayshore in this August 5, 1965 photo. The consist includes three single level coaches on the head-end followed by the newer bi-level cars on this foggy day in the Bay Area—August 5, 1965. *Photo by Friedrich, William S. Kuba Collection*

Fog can be a common element on the West Coast as can be observed on this April 1968 spring day. The nine-car train has a mixture of single level two-tone grey coaches not far from the San Francisco Station. *Photo by Friedrich, William S. Kuba Collection*

Train 130 is departing the San Francisco station with a consist of the heavy weight single level coaches. It is September 18, 1967, and the SP operated but one intercity train in and out of San Francisco that was listed in the timetable as handling commuter passengers, the Coast Daylight. *Photo by Friedrich, William S. Kuba Collection*

The SP also operated GP9s, such as the 3007 shown here, with steam generators in commuter train service. The consist for train 147, en route to San Francisco, is three cars including two bi-level cars and one single level car carrying the markers. No. 147 is pausing at Redwood City and the passengers are heading for the train in the August 1972 photo. *William S. Kuba Collection*

GP9 No. 3194 is also equipped with a steam generator, but has dynamic brakes in addition. It is posed here ready to depart San Francisco with train No. 120 for San Jose. No. 120 had a three-car consist, heavy weight coach 3194 and two bi-levels, 3719 and 3736. It is the day after Christmas, December 26, 1978. *Patrick C. Dorin*

One can expect just about any kind of combination of motive power and passenger equipment (as well as freight and work cars) in the railroad industry. This scene at San Francisco, September 1978, shows an Amtrak General Electric P30, No. 700, ready to handle a consist of heavy weight SP commuter cars. *Photo by Bryan Griebenow, William S. Kuba Collection*

Amtrak No. 712 is shown here with two bi-level coaches for the consist in this fall 1978 portrait at San Francisco. There is additional Amtrak power behind the bi-level cars. *Photo by Bryan Griebenow, William S. Kuba Collection*

This December 1978 scene illustrates the variety and mixture of passenger equipment being pre-pared for the rush hour services later in the day. As one can see, the single level cars are coupled to the bi-levels on at least two tracks in this scene. It would be interesting if a model railroad producer made the SP single level heavy weight coaches which could be mixed with the Walthers HO gauge model of the SP bi-levels. *Patrick C. Dorin*

The SP equipment did not have the head-end power arrangement such as could be found on commuter train equipment in the east and midwest, as well as the newer Amtrak passenger cars. Note the three hoses connected between the locomotive and the passenger car in this photo: the steam line for heating, air signal line and the air brake line. *Patrick C. Dorin*

This end view (December 1978) illustrates the couplers and the placement of the steam pipe, and the two air lines on coach 2127. Again we have a mix of equipment, which has the blue flag attached on the consist at the San Francisco station. *Patrick C. Dorin*

This side view of coach 2109 illustrates the window placement on this group of commuter coaches (December 1978). Note the lettering for the number with SP initials followed by the number. This will be a lead car on an outbound rush hour train from San Francisco. *Patrick C. Dorin*

The first group of SP bi-level coaches came in the two-tone grey color scheme. The cars had sub-lettering "Commute" at each end of the car between the two levels. The 3705 was photographed at San Francisco in 1955. *Wilbur C. Whittaker*

The second order of SP bi-level coaches, gallery cars to be more precise, came with five windows in each quarter section as illustrated here with SP 3745. The cars were painted in the full grey colors instead of the two-tone variety. *Pullman-Standard, John Sward Collection*

Interior view of the upper level of the SP gallery commuter cars. *Pullman-Standard, John Sward Collection*

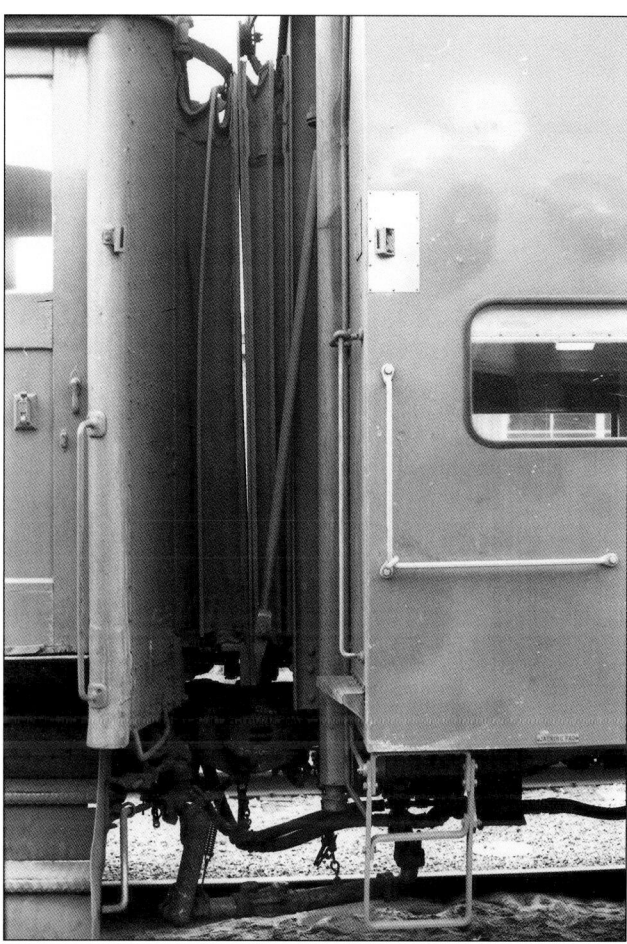

The connection between a bi-level car and a single level heavy weight commuter car. *Patrick C. Dorin*

The newest bi-level cars were also steam heated as this photo shows the diaphragms, and the steam lines at the bottom. *Patrick C. Dorin*

We will finish up the chapter on the Southern Pacific commuter services by taking a look at some of the trains that once operated in what is "NOW" commuter train territory on the former SP (now UP) lines. This photo shows the Lark, an overnight Los Angeles–San Francisco train. Train 76 has arrived at Los Angeles after operating over the future Caltrain route between San Francisco, San Jose and Gilroy at the north end, and over the Metrolink Ventura County Line at the south end. *R. S. Plummer, William S. Kuba Collection*

Train No. 2, the Sunset Limited, operated east out of Los Angeles over part of the Metrolink territory. *Photo by Friedrich, William S. Kuba Collection*

Train 102, the City of San Francisco (Chicago–Oakland service) is shown here at Sacramento in July 1966. This is the route of Amtrak California's Capitol Corridor line to and from Oakland and San Jose. *William S. Kuba Collection*

The last passenger motive power purchased by the Southern Pacific were the SDP-45s, equipped with steam generators for passenger service. The 3204 is shown here with a five-car train at Martinez. *Photo by Friedrich, William S. Kuba Collection*

Train No. 120, en route to San Jose, is shown here at South San Francisco with a two-car consist in February 1969. *Erik Frodsham*

This three-car train with both single level and bi-level coaches was caught on film at San Jose on April 2, 1985. The SD units for power were the 4450 and 4451. *Erik Frodsham*

This top-down photo illustrates the SP trackage at San Jose dating back to January 1985. The 3195 is leading the three-car consist of train 26 to San Jose. *Erik Frodsham*

The newest passenger power on the SP, such as the 3201 (SDP-45), is heading up a consist of single level commuter coaches on train No. 43 bound for San Francisco. This view was taken at San Jose in April 1985. *Erik Frodsham*

This RDC car on the SP did not operate in the commuter train services, but did provide a local service north of the Bay Area to Eureka, California. The car was numbered SP-10. These two views illustrate the modifications made to this RDC-1, such as the train number boards above each end, and different door systems at the right end of the car in these two photos. The side view was taken at Eureka in July 1966, while the end view was taken at Willits in April 1971. *Erik Frodsham*

We will finish up the photo section of the Southern Pacific Commuter Train Services with two illustrations showing the rear of two sets of consists. The single level group has the 2148 as the last car, while the bi-level cars complete the consist with the 3705. *Erik Frodsham*

It is a beautiful day in April 2003 as train No. 648 arrives at San Diego with the 2103 for power. The skyline and the palm trees add to the beauty on this 10th day of April. *Steve Glischinski*

Chapter 2
THE COASTER: SAN DIEGO

The Coaster is the name for the commuter train service operating between San Diego and Oceanside, a distance of 41.1 miles. The service operates over the San Diego Northern Railway, which is a subsidiary of the North County Transit District. This rail line was once part of the Santa Fe System.

The historic start date for the new train service was February 27, 1995. Coaster serves eight stations: San Diego, Old Town, Sorrento Valley, Solana Beach, Encinitas, Carlsbad Poinsettia, Carlsbad Village and Oceanside. Oceanside is also a connection with the southern end of the Metrolink Commuter Rail Service from Los Angeles. Thus it can be said, that Oceanside is served by three rail passenger carriers, the Coaster, Metrolink and Amtrak. Very few center cities can claim that as we move through the 21st Century, let alone one in the outer reaches of the suburban territory.

The Coaster operates two coach yards, one at San Diego and one just to the north of Oceanside. The Oceanside yard also provides space for the Metrolink trains from Los Angeles.

The train service has been expanded over the years. As of 2007, as this is being written, the Coaster operates 11 trains in each direction Monday through Thursday. Two additional trains are operated in each direction on Friday evenings April through October. This compares with 9 trains in each direction in the year 2000 without the Friday evening trains. And that isn't all—when the Padres baseball team is playing any evening Monday through Thursday at Petco Park, there is an additional evening train in each direction. The Coaster operates four trains in each direction on Saturday, which was also the case in the year 2000.

The Coaster service, along with the Amtrak

Surfliner train frequency to San Diego, has provided the largest number of passenger trains serving San Diego in its entire history. In 2007 Amtrak is currently running 11 trains each way daily and 12 each way on Fridays, Saturdays and Sundays. With the Coaster's frequency of a minimum of 11 each way on Monday through Thursday, San Diego has at least 44 trains per day. Of course, the frequency drops substantially on Saturdays and Sundays, but even with that, San Diego has never had 12 Amtrak trains to and from Los Angeles Friday through Sunday.

Ridership has continued to grow since the start of operations in 1995. The first year ridership in 1995 was over 179,000. The weekday ridership reached the 4,900 to 5,000 range by 2003, and as this is being written in early 2007, the number is 5,645. The total number of passengers carried for the fiscal year 2006 was over 1.5 million.

The Coaster and Amtrak services for Southern California has had a positive effect on the economic development in the area, in addition to adding a new sense of community for many people. It is another case that demonstrates that rail passenger service has far more benefits than simply the passenger revenues from ticket sales.

Year	Annual Ridership	Weekday Ridership
1995	179,378 *	Not Available through 2000
1996	736,776	
1997	909,976	
1998	1,031,268	
1999	1,240,225	
2000	1,187,749	
2001	1,206,839	4455
2002	1,281,444	4775
2003	1,348,453	4950
2004	1,429,020	5309
2005	1,432,468	5294
2006	1,554,450	5778

* Reflects the ridership for five months in 1995.

Motive Power and Equipment

The Coaster trains are powered by a fleet of F40PH-2Cs and F59PHIs. The passenger equipment was built by Bombardier. The fleet contains bi-level coaches and cab cars for the push-pull operation. The coach cars can seat 135 passengers while the cab cars provide seating for 130 passengers.

Motive Power Roster

Type	Number Series	Horsepower	Date	Builder
F40PH-2C	2101 to 2105	3,000	1994	Morrison-Knudsen
F59PHI	3001 to 3002	3,000	2001	Electro-Motive Div.

Passenger Equipment Roster

Type	Number Series	Builder	Date
Cab Coaches	2301 to 2308	Bombardier	1994-5
Cab Coaches	2309, 2310	Bombardier	2003
Trailer Coaches	2201 to 2208	Bombardier	1994-5
	2401 to 2406	Bombardier	1998
	2501 to 2504	Bombardier	2003

The following photos and tables illustrate the Coaster Services as we move through the first decade of the 21ˢᵗ Century.

The Coaster timetable to the right was issued in November 2007, and illustrates the frequency and the time frame for the train service. The schedule shows that there were five rush hour trains to San Diego in the morning, with an equal number to Oceanside in the evening. Friday evening trains during the summer of 2007 provided a new level of convenience for passengers to and from San Diego for many recreational activities. Note the Monday–Thursday evening train, No. 659, operated for the Padres at Petco Park in the summer 2007 schedule.

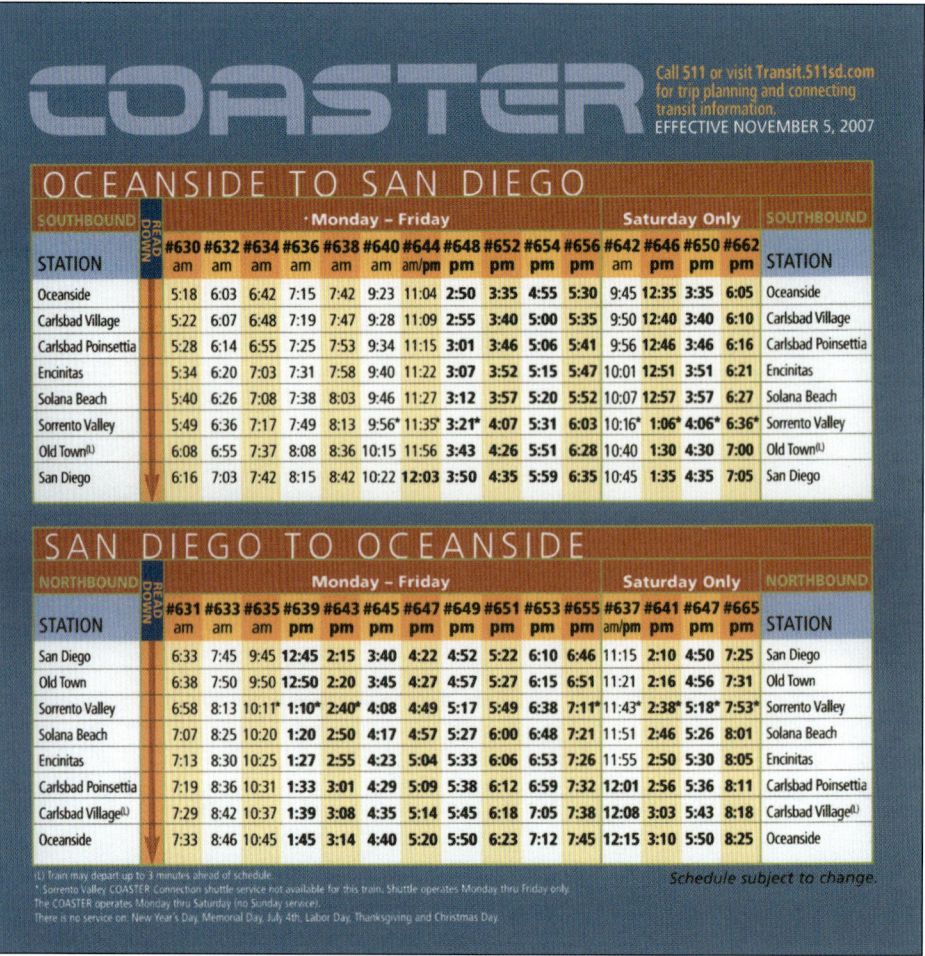

COASTER — Call 511 or visit Transit.511sd.com for trip planning and connecting transit information. EFFECTIVE NOVEMBER 5, 2007

OCEANSIDE TO SAN DIEGO — SOUTHBOUND (READ DOWN)

STATION	#630 am	#632 am	#634 am	#636 am	#638 am	#640 am	#644 am/pm	#648 pm	#652 pm	#654 pm	#656 pm	#642 am	#646 pm	#650 pm	#662 pm	STATION
	Monday – Friday											Saturday Only				
Oceanside	5:18	6:03	6:42	7:15	7:42	9:23	11:04	2:50	3:35	4:55	5:30	9:45	12:35	3:35	6:05	Oceanside
Carlsbad Village	5:22	6:07	6:48	7:19	7:47	9:28	11:09	2:55	3:40	5:00	5:35	9:50	12:40	3:40	6:10	Carlsbad Village
Carlsbad Poinsettia	5:28	6:14	6:55	7:25	7:53	9:34	11:15	3:01	3:46	5:06	5:41	9:56	12:46	3:46	6:16	Carlsbad Poinsettia
Encinitas	5:34	6:20	7:03	7:31	7:58	9:40	11:22	3:07	3:52	5:15	5:47	10:01	12:51	3:51	6:21	Encinitas
Solana Beach	5:40	6:26	7:08	7:38	8:03	9:46	11:27	3:12	3:57	5:20	5:52	10:07	12:57	3:57	6:27	Solana Beach
Sorrento Valley	5:49	6:36	7:17	7:49	8:13	9:56*	11:35*	3:21*	4:07	5:31	6:03	10:16*	1:06*	4:06*	6:36*	Sorrento Valley
Old Town (L)	6:08	6:55	7:37	8:08	8:36	10:15	11:56	3:43	4:26	5:51	6:28	10:40	1:30	4:30	7:00	Old Town (L)
San Diego	6:16	7:03	7:42	8:15	8:42	10:22	12:03	3:50	4:35	5:59	6:35	10:45	1:35	4:35	7:05	San Diego

SAN DIEGO TO OCEANSIDE — NORTHBOUND (READ DOWN)

STATION	#631 am	#633 am	#635 am	#639 pm	#643 pm	#645 pm	#647 pm	#649 pm	#651 pm	#653 pm	#655 pm	#637 am/pm	#641 pm	#647 pm	#665 pm	STATION
	Monday – Friday											Saturday Only				
San Diego	6:33	7:45	9:45	12:45	2:15	3:40	4:22	4:52	5:22	6:10	6:46	11:15	2:10	4:50	7:25	San Diego
Old Town	6:38	7:50	9:50	12:50	2:20	3:45	4:27	4:57	5:27	6:15	6:51	11:21	2:16	4:56	7:31	Old Town
Sorrento Valley	6:58	8:13	10:11*	1:10*	2:40*	4:08	4:49	5:17	5:49	6:38	7:11*	11:43*	2:38*	5:18*	7:53*	Sorrento Valley
Solana Beach	7:07	8:25	10:20	1:20	2:50	4:17	4:57	5:27	6:00	6:48	7:21	11:51	2:46	5:26	8:01	Solana Beach
Encinitas	7:13	8:30	10:25	1:27	2:55	4:23	5:04	5:33	6:06	6:53	7:26	11:55	2:50	5:30	8:05	Encinitas
Carlsbad Poinsettia	7:19	8:36	10:31	1:33	3:01	4:29	5:09	5:38	6:12	6:59	7:32	12:01	2:56	5:36	8:11	Carlsbad Poinsettia
Carlsbad Village (L)	7:29	8:42	10:37	1:39	3:08	4:35	5:14	5:45	6:18	7:05	7:38	12:08	3:03	5:43	8:18	Carlsbad Village (L)
Oceanside	7:33	8:46	10:45	1:45	3:14	4:40	5:20	5:50	6:23	7:12	7:45	12:15	3:10	5:50	8:25	Oceanside

(L) Train may depart up to 3 minutes ahead of schedule.
* Sorrento Valley COASTER Connection shuttle service not available for this train. Shuttle operates Monday thru Friday only.
The COASTER operates Monday thru Saturday (no Sunday service).
There is no service on: New Year's Day, Memorial Day, July 4th, Labor Day, Thanksgiving and Christmas Day.

Schedule subject to change.

The summer 2007 schedule below is without the evening train schedules as compared with the November 2007 schedule. The Coaster website: *www.gonctd.com*

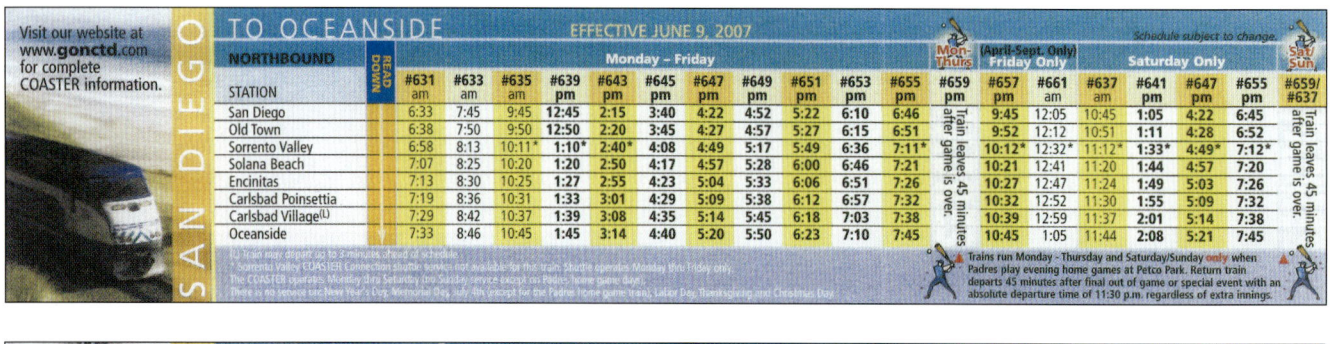

Visit our website at **www.gonctd.com** for complete COASTER information.

TO OCEANSIDE — NORTHBOUND — EFFECTIVE JUNE 9, 2007 (READ DOWN)

STATION	#631 am	#633 am	#635 am	#639 pm	#643 pm	#645 pm	#647 pm	#649 pm	#651 pm	#653 pm	#655 pm	Mon-Thurs #659 pm	Friday Only #657 pm	Friday Only #661 am	Saturday Only #637 am	Saturday Only #641 pm	Saturday Only #647 pm	Saturday Only #655 pm	Sat/Sun #659/#637
San Diego	6:33	7:45	9:45	12:45	2:15	3:40	4:22	4:52	5:22	6:10	6:46	9:45	12:05	10:45	1:05	4:22	6:45		
Old Town	6:38	7:50	9:50	12:50	2:20	3:45	4:27	4:57	5:27	6:15	6:51	9:52	12:12	10:51	1:11	4:28	6:52		
Sorrento Valley	6:58	8:13	10:11*	1:10*	2:40*	4:08	4:49	5:17	5:49	6:36	7:11*	10:12*	12:32*	11:12*	1:33*	4:49*	7:12*		
Solana Beach	7:07	8:25	10:20	1:20	2:50	4:17	4:57	5:28	6:00	6:48	7:21	10:27	12:41	11:20	1:44	4:57	7:20		
Encinitas	7:13	8:30	10:25	1:27	2:55	4:23	5:04	5:33	6:06	6:51	7:26	10:27	12:47	11:24	1:49	5:03	7:26		
Carlsbad Poinsettia	7:19	8:36	10:31	1:33	3:01	4:29	5:09	5:38	6:12	6:57	7:32	10:32	12:52	11:30	1:55	5:09	7:32		
Carlsbad Village (L)	7:29	8:42	10:37	1:39	3:08	4:35	5:14	5:45	6:18	7:03	7:38	10:39	12:59	11:37	2:01	5:14	7:38		
Oceanside	7:33	8:46	10:45	1:45	3:14	4:40	5:20	5:50	6:23	7:10	7:45	10:45	1:05	11:44	2:08	5:21	7:45		

Train leaves 45 minutes after game is over. (Mon-Thurs) (April-Sept. Only Friday) Train leaves 45 minutes after game is over. (Sat/Sun)

▲ Trains run Monday - Thursday and Saturday/Sunday only when Padres play evening home games at Petco Park. Return train departs 45 minutes after final out of game or special event with an absolute departure time of 11:30 p.m. regardless of extra innings.

Schedule subject to change.

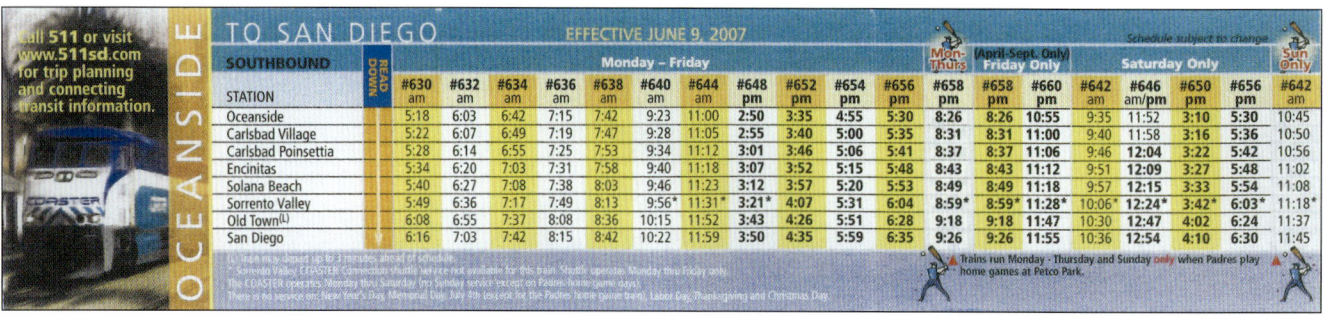

Call 511 or visit **www.511sd.com** for trip planning and connecting transit information.

TO SAN DIEGO — SOUTHBOUND — EFFECTIVE JUNE 9, 2007 (READ DOWN)

STATION	#630 am	#632 am	#634 am	#636 am	#638 am	#640 am	#644 am	#648 pm	#652 pm	#654 pm	#656 pm	Mon-Thurs #658 pm	Friday Only #658 pm	Friday Only #660 am	Saturday Only #642 am	Saturday Only #646 am/pm	Saturday Only #650 pm	Saturday Only #656 pm	Sun Only #642 am
Oceanside	5:18	6:03	6:42	7:15	7:42	9:23	11:00	2:50	3:35	4:55	5:30	8:26	8:26	10:55	9:35	11:52	3:10	5:30	10:45
Carlsbad Village	5:22	6:07	6:49	7:19	7:47	9:28	11:05	2:55	3:40	5:00	5:35	8:31	8:31	11:00	9:40	11:58	3:16	5:36	10:50
Carlsbad Poinsettia	5:28	6:14	6:55	7:25	7:53	9:34	11:12	3:01	3:46	5:06	5:41	8:37	8:37	11:06	9:46	12:04	3:22	5:42	10:56
Encinitas	5:34	6:20	7:03	7:31	7:58	9:40	11:18	3:07	3:52	5:15	5:48	8:43	8:43	11:12	9:51	12:09	3:27	5:48	11:02
Solana Beach	5:40	6:27	7:08	7:38	8:03	9:46	11:23	3:12	3:57	5:20	5:53	8:49	8:49	11:18	9:57	12:15	3:33	5:54	11:08
Sorrento Valley	5:49	6:36	7:17	7:49	8:13	9:56*	11:31*	3:21*	4:07	5:31	6:04	8:59*	8:59*	11:28*	10:06*	12:24*	3:42*	6:03*	11:18*
Old Town (L)	6:08	6:55	7:37	8:08	8:36	10:15	11:52	3:43	4:26	5:51	6:28	9:18	9:18	11:47	10:30	12:47	4:02	6:24	11:37
San Diego	6:16	7:03	7:42	8:15	8:42	10:22	11:59	3:50	4:35	5:59	6:35	9:26	9:26	11:55	10:36	12:54	4:10	6:30	11:45

▲ Trains run Monday - Thursday and Sunday only when Padres play home games at Petco Park.

Schedule subject to change.

The 2102 is leading a train pausing at Carlsbad, California, in March 2002. *Robert Larson*

The commuter rail route between San Diego and Oceanside has an unlimited amount of beautiful scenery. What passengers can enjoy and relax is shown here by this photo in April 2003. Train 639 is rolling through the Miramar Hill area with the F59PHI, No. 3001, for power. *Steve Glischinski*

Train No. 656 is powered by the 3001 as it rolls over the bridge near Cardiff, California. The four-car train was the last departure of the day (at 5:28 p.m.) from Oceanside to San Diego when this photo was taken in April 2003. *Steve Glischinski*

Train No. 649 is a rush hour train from San Diego to Oceanside with a five-car consist on this April 15, 2003. No. 649 is near Solana Beach, California. It has about 20 minutes left on the schedule to Oceanside. *Steve Glischinski*

This view shows the shop area for the Coaster, the North County Transportation District, at Oceanside. The 2105 is ahead of still another locomotive and one coach in this view, which was taken on March 5, 2005. *Robert Larson*

The Coaster trains operate in the push-pull mode, and train No. 648 is at the Del Mar, California, depot with a mid-afternoon schedule from Oceanside to San Diego. The F59PHI No. 3001 is doing the honors, pushing the train on this April 2003 afternoon. *Steve Glischinski*

F40PHC No. 2105 was part of the group purchased from Morrison-Knudson in 1994. The 3,000-horsepower engines provide adequate power for the passenger train services. The 2105's portrait was taken on December 26, 1995. *William S. Kuba*

This view illustrates the right side of an F40PHC, No. 2104, part of the 2101 to 2105 series. *North County Transit District*

This F59PHI, No. 3002, is heading up a five-car train for this publicity photo for the Coaster Services by North County Transit District.

Metrolink serves a highly populated area in Southern California. Yet the routes pass through some very scenic areas, such as this four-car train rolling over the Bridge. The FP59H, No. 872, is providing the power for the four-car train, which is in the push mode as the headlights are not on. *Metrolink Photo*

Chapter 3
THE METROLINK SYSTEM: LOS ANGELES

Who would have thought in 1982, that ten years later in 1992; a new commuter rail service would begin operation in the Los Angeles Metropolitan Area with a system of eventually seven routes, including a belt line between suburbs. As of this writing in 2007, there are only six commuter rail systems that operate seven or more routes: Long Island in New York; Southeastern Pennsylvania Transportation Authority in Philadelphia; Metra in Chicago; the Massachusetts Bay Transportation Authority, known as the MBTA; New Jersey Transit; and GO Transit in Toronto. Metrolink is administered by the Southern California Regional Rail Authority, with the initials SCRRA. What is the Metrolink System and where

does it operate?

October 26, 1992, should be listed as a California State Historic Holiday. On that special day, the Metrolink began commuter train service in the Los Angeles Area. The first three routes started 16 years after a 1976 California legislative action to establish county level transportation commissions in four of the Los Angeles Area counties. This in turn led to additional planning, initially for a 150-mile commuter rail system in 1980. However, financing and other plans did not really get launched until the late 1980s, which led to more action to get "Things on Track." Just one example was the Southern Pacific's offer to sell some of its under-utilized freight lines in the LA area. Part of this involved a route between Los Angeles and San Bernardino, and routes on the SP's Valley and Coast Lines to Saugus and Moorpark.

The year 1991 produced still another milestone for the commuter rail planning, one that called for a 412-mile system serving the five-county area. The new plan identified the seven routes linking Los Angeles with Moorpark, Santa Clarita, San Bernardino, Oceanside and Riverside. Still another route would link San Bernardino, Riverside and Irvine. Still another route was considered between San Bernardino and Redlands. (This latter route is now under consideration.)

An agreement was reached with the Union Pacific in early 1992 for a route between Los Angeles and San Bernardino. Additional agreements were reached with the Santa Fe for the routes serving points in Orange, Riverside and San Bernardino counties. This also involved trackage rights over the Santa Fe between Los Angeles and San Bernardino via Fullerton and Riverside. The route toward Oceanside on the coast was also part of the work along with additional cooperative work with the San Diego County transportation agencies, which were working on the establishment of the Coaster Commuter service between Oceanside and San Diego.

The train service began with 23 locomotives, F59PH, numbered 851 to 873, which were built in 1992 and '93. There were 94 bi-level coaches of which 31 were cab cars and 63 were trailer coaches, numbered 601 to 631 and 101 to 163 respectively. See the roster later in this chapter for more information.

The first three Metrolink routes, which began operating on October 26, 1992, included:

Route	Rail Line	Mileage	Original Train Frequency
Los Angeles to Moorpark	SP Coast Line	47	4 Each Way
Los Angeles to Santa Clarita	SP Valley Line	35	3 Each Way
Los Angeles to Pomona	Ex-SP State Street/ Baldwin Park Branches Now SCRRA	31	5 Each Way

The train frequencies listed were Rush Hour Only.

The Next Steps were the Route Extensions in 1993 and 1994

1993
Los Angeles–San Bernardino	SCRRA	51	3 Each Way Rush Hour
Los Angeles–Riverside	UP	57	3 Each Way Rush Hour
Los Angeles–Oceanside	Santa Fe	87	3 Each Way Rush Hour

December 1994
San Bernardino–Riverside Irvine	Santa Fe	63	2 Each Way Rush Hour

The FP59H, No. 860, was part of the first group of motive power to be purchased for the Metrolink services in 1992. This unit displays the original paint scheme applied to the motive power with much more blue on the nose and cab. *Metrolink Photo*

The year 1992 was the initial mile post as the growth rate continued with new routes going into service, the double tracking of many segments of trackage, and expanded train frequencies on all of the routes. Moving through the mid to late 1990s and into the 21st Century, the Los Angeles Area has a train service level that never existed before in the State of California—especially with its dependence on the expressways. What people thought was freedom to go anywhere, actually meant little freedom as one sits and crawls on the road system throughout the area.

The 21st century model of Metrolink has seven routes in operation with a solid train service. The routes include the following:

Route	End Terminals	Mileage
San Bernardino	San Bernardino–LA Union Station via Claremont	56.5
Antelope Valley	Lancaster–LA Union Station	76.6
Ventura County	Montalvo–LA Union Station	70.9
Orange County	Oceanside–LA Union Station	87.2
Inland Empire/Orange County	San Bernardino–Oceanside via Riverside and Irvine	100.1
91 Line	San Bernardino–LA Union Station via Riverside and Fullerton	61.6
Riverside	Riverside–LA Union Station via East Ontario and Industry	59.1

The Los Angeles facility services the motive power between runs. This color photo shows a picturesque cloudy red sky reflecting on four engines. Metrolink, as of the late 1990s and early 2000s, had two types of passenger power, the FP59H and the F40PHI. *Metrolink Photo*

The total route mileage for the seven routes is 512 miles. However, this total includes the shared mileage of several routes as they converge on Los Angeles. Also the Inland Empire / Orange County route shares the routes to San Bernardino and to Oceanside. The total system mileage is 388 miles.

The following tables list the number of trains to and from Los Angeles for each route during the rush hour and non-rush hour periods as this is being written in 2007.

The rush hour arrivals in Los Angeles are counted between 5:30 a.m. and 9:00 a.m. Rush hour departures from LA are counted from 3:30 p.m. to 6:00 p.m. Train counts are based on the train number series for the route. The following numbers do not include the Amtrak Rail 2 Rail schedules, which follow this table of train service levels.

Route	Weekday Rush Hour	Weekday Non-Rush Hour	Sat.	Sun.
Ventura County Line				
100 Series Train Numbers				
To Los Angeles	4	7	-	-
From Los Angeles	4	7	-	-
Antelope County Line				
200 Series Train Numbers				
To Los Angeles	5	7	6	3
From Los Angeles	4	8	6	3
San Bernardino				
300 Series Train Numbers				
To Los Angeles	8	9	10	7
From Los Angeles	7	10	10	7
Riverside Line				
400 Series Train Numbers				
To Los Angeles	4	2	-	-
From Los Angeles	3	3	-	-
Orange County Line				
600 Series Train Numbers				
To Los Angeles	6	4	4	4
From Los Angeles	4	5	4	4
91 Line				
700 Series Trains—Riverside				
To Los Angeles	2	2	-	-
From Los Angeles	2	3	-	-

Inland Empire–Orange County

The IEOC Line consists of train numbers in the 800 series for the route, but also includes the 600 series trains from the route to and from Los Angeles. The IEOC line is a belt line commuter route. The weekday schedule includes one rush hour train in each direction from the entire distance between San Bernardino and Oceanside, plus two trains each way between Oceanside and Riverside. The total number of weekday westbound trains toward Irvine and Oceanside is 17. The eastbound total toward Riverside and San Bernardino is 18. These trains only operate over part of the route.

The weekend services include seven trains each way on Saturday and six each way on Sunday. There are two trains each way between San Bernardino and Oceanside on the weekend, with one additional train each way between Riverside and Oceanside on Saturday.

One of the interesting aspects of the new commuter services on routes to Oceanside and Ventura County is the cross-ticketing arrangements between Amtrak and Metrolink. It is called Rail 2 Rail, and Metrolink passengers with multiple ride tickets have access to the Amtrak trains that stop at the various stations. This system increases the availability of train services. It is also very similar to the pooling of passenger services once practiced by the Great Northern and the Northern Pacific with the Union Pacific on the Seattle–Portland corridor, and the Soo Line between Duluth-Superior and the Twin Cities. This is one reason why, when studies are done for new services, it certainly pays to review history of past train services and operations.

The Amtrak Rail 2 Rail weekday and weekend count is as follows:

Route	Weekday	Weekend
Ventura County Line (LA–Montalvo)		
To Los Angeles	4	5
From Los Angeles	4	5
Ocean County Line (LA–Oceanside)		
To Los Angeles	11	12
From Los Angeles	11	12

For more information, and the latest updates and schedules, one can check the Metrolink website at *www.metrolinktrains.com*

Depending upon the time of day and the train schedule, passenger consists will vary to accommodate the passenger loads. This three-car non-rush hour train can handle over 400 passengers, and it gives one the idea of the flexibility and effectiveness of this type of bi-level passenger equipment. *Metrolink Photo*

MOTIVE POWER AND PASSENGER EQUIPMENT

The motive power fleet for Metrolink has grown since the initial purchase of diesel power in 1992. As of 2006, the following is the roster of Metrolink motive power in operation on the seven routes:

Locomotive Roster

Type	Number Series	Built
F40PHI	800	2004
F59PH	851 to 869	1992
F59PH	870 to 873	1993
F59PHI	874 to 881	1995
F59PHI	882 to 883	1996
F59PHI	884 to 887	2001

Through 2006, Metrolink had 38 diesel units plus one unit leased from Sound Transit in 2004 for a total of 39 locomotives.

As of late 2007, there are 15 units (MP36PH-3C) on order from Motive Power Incorporated.

All of the F40 and F59 units have 3,000 horsepower, while the MP36s have 3,600 horsepower. .

Passenger Roster

All of the passenger cars operated by Metrolink were built by Bombardier or its predecessor. The AAR designation / reporting marks are SCAX. As of late 2007, Metrolink has 33 Cab Cars plus 4 leased from Sound Transit, and 2 from Altamont Commuter Express. The total number of Coaches is 116 with 8 leased from Sound Transit and 2 from Altamont Commuter Express. Metrolink has 53 Cab Cars and 54 Coaches on order from the Roten Company in South Korea as of late 2007.

Type	Number Series	Built
Cab Coaches	601 to 631	1992
	632 to 637	1997-98
Trailer Coaches	101 to 163	1992-93
	164 to 182	1997-98
	183	
	185 to 210	2002

This view peaks into the interior of the lower level of the Bombardier bi-level coach. The window is reflecting the sunlight, and it is apparent that the passengers are relaxed and ready for the trip. *Metrolink Photo*

The Metrolink station facilities are inviting, and this appearance and atmosphere add more reasons for passengers to become repeat travelers. This view shows the entry to the Claremont station on the San Bernardino line. *Metrolink Photo*

This writer had an opportunity to ride the Metrolink during the summer of 2005. Many people told me that they do not know what they would do without the train service. A few people expressed the opinion that additional bus routes and perhaps new light-rail services are needed to connect with the Metrolink trains at various stations because of the urban sprawl. One person expressed the thought that the only time he rides a bus is either to catch a train or a plane.

A nurse, now living in Minnesota, said she had once lived in the LA area and at the beginning of her career, she could drive to work in 20 minutes. As time went by, it was taking her over 2 hours to drive one way to the hospital—just sitting on the expressway. She referred to it as the "Elongated Parking Ramp" for gasoline consumption. She moved to Minnesota for a better environment from the driving. She expressed the opinion that she now sees the same problems in the Twin Cities area. She added with a smile, "Minnesota has to follow California and get on track."

Two other women told this writer that they met their future husbands on the Metrolink trains. One added, also with a smile, "Meeting my husband on the train has kept our marriage on track."

The average weekday ridership was over 43,400 passengers in mid-2007.

There are future plans for additional service levels, which will require more track and signal work. For

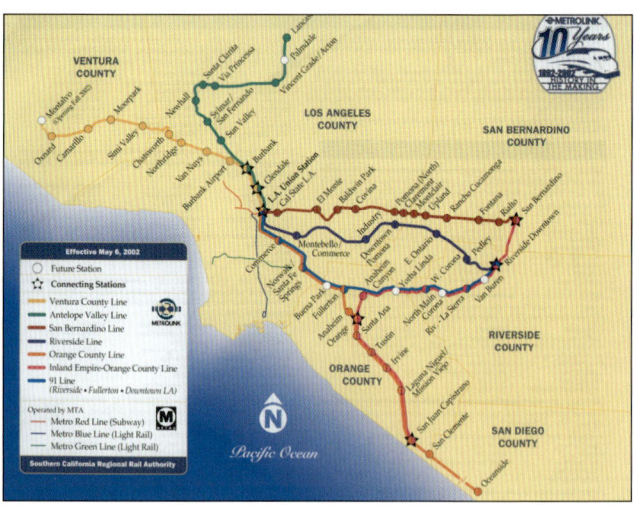

2002 map showing Metrolink routes

example, the Redlands–San Bernardino route, a 9-mile extension, will be serving seven or eight stations. Diesel Multiple Unit equipment is under consideration to provide a connecting service to the through trains at the downtown San Bernardino station. An extension of the 91 Line to connect Riverside and Perris, a distance of 22 miles, is also in the planning stage. The extended line will serve five stations.

Fortunately, California has had the common sense to see that rail services are the cure for congestion, pollution and other economic problems.

The map below illustrates the Metrolink commuter train service routes as of early 2008.

Eastbound train No. 308 is departing the Claremont depot in this June 2005 photo. Note the yellow lines on the station platforms to keep passengers alert from getting too close to the tracks. *Patrick C. Dorin*

This is the El Monte depot, which is on the San Bernardino line. It is a little over 20 minutes from downtown Los Angeles, train time that is. What more could one ask for with the palm trees as part of the station décor. *Metrolink Photo*

Rialto is a short distance from San Bernardino, and has been built to reflect the historic patterns of many of the railroad stations. The station has been named the John Longville Depot. This scene illustrates the entry from the street. *Metrolink Photo*

The Los Angeles Station has stub end tracks. The trains arrive in the push mode and move up to the end of the track as illustrated here in June 2005. *Patrick C. Dorin*

This view (June 2005) shows a train ready to depart Los Angeles with the motive power in pull mode. The loading and unloading platforms have been reconditioned and/or rebuilt as LA now handles more passenger trains than ever. *Patrick C. Dorin*

As interesting concept about Los Angeles is that the Metrolink commuter trains and the Metro Rail light-rail train arrive at the same station for convenient cross-the-platform transfers. Note the overhead electric wire over the track to the right. The three-car Metrolink train to the left has two units for power on this particular afternoon in June 2005. *Patrick C. Dorin*

This end view of cab cars, such as the 616 to the right, shows only one window in the engineer's compartment on the right side of the car. This view at the LA station shows one of the center tracks between the tracks at the platforms. *Patrick C. Dorin*

Cab car No. 636 has two forward windows. *Patrick C. Dorin*

A train is getting ready to depart the Los Angeles station with a consist of four cars on this pleasant early summer afternoon in June 2005. *Patrick C. Dorin*

This ¾ view (June 2005) illustrates the color scheme application of the striping on the coach sides as well as on the front of the cab car. *Patrick C. Dorin*

A close-up view showing the color scheme and how it matches from car to car in the train consists. *Patrick C. Dorin*

Coach No. 177.
Patrick C. Dorin

F59PH No. 871 in the current blue-and-white color scheme. Note the placement of the Metrolink insignia and wording on both the nose and the side of the motive power. *Patrick C. Dorin*

Motive power line up of the 875, a F59PHI, and the 871 ready for the next departures out of the Los Angeles station (June 2005). *Patrick C. Dorin*

A five-car train (March 2005) is laying over at Oceanside, which is the connecting point with the Coaster for the commuter trains to and from San Diego. As mentioned earlier, Oceanside is one of very few communities in North America served by three passenger railroads. *Robert Larson*

Train No. 686 is arriving at the Santa Ana, California station August 2005. This top-down photo illustrates the high quality of the station platforms. *Steve Glischinski*

Train No. 806 is departing Santa Ana for Los Angeles with F59PH No. 865 for power and a four-car consist in August 2005. *Steve Glischinski*

The Amtrak services between Los Angeles and San Diego were once operated with Amfleet passenger equipment. This photo (March 1993) illustrates push-pull operation for the trains with the cab car at San Diego. *Steve Glischinski*

Chapter 4
AMTRAK'S SURFLINERS: RAIL 2 RAIL

Amtrak California has a positive working relationship with the Coaster and the Metrolink commuter railroads. Amtrak provides train services for Metrolink and Coaster ticket holders—and the slogan for the service is:

RAIL 2 RAIL. AN EASIER WAY TO GO.

Amtrak California offers train service for Metrolink monthly pass ticket holders on two of the Metrolink routes: the Orange County line between Los Angeles and Oceanside, and the Ventura County line between Montalvo and Los Angeles. The Orange County and Ventura County timetables list the schedules of the Amtrak trains serving the same stations as Metrolink.

Passengers with Coaster monthly tickets can use Amtrak trains at San Diego, Solana Beach and Oceanside. Amtrak ticket holders can use either of the train services for those designated stations.

The Amtrak Surfliner trains have been reequipped with the new bi-level Surfliner equipment including coaches for unreserved seating, Business Class Service, and Café Cars. The trains operate with the push-pull concept used by the Coaster and Metrolink trains. There is no doubt about it, Amtrak California is demonstrating to North America the value, convenience and new levels of benefits with and for rail passenger transportation.

The following timetables and photos illustrate the coordinated services between Amtrak and the Commuter Rail Services.

AMTRAK

PACIFIC SURFLINER® NORTHBOUND
SAN DIEGO–LOS ANGELES–SAN LUIS OBISPO

Effective JANUARY 21, 2008

San Diego • Orange County • Los Angeles • Santa Barbara • San Luis Obispo • Paso Robles

Train Number ▶	Mile		799	763	565	567	769	571	573	775	577	579	583	785	587	589	591	595	597
Normal Days of Operation ▶			Daily	Daily	Daily	Mo-Fr	Daily	SaSu	Mo-Fr	Daily	SaSu	Mo-Fr	Daily	Daily	SaSu	Mo-Fr	SaSu	Daily	FrSaSu
San Diego, CA (Tijuana) (PT)	0	Dp		6 10A	7 05A	8 10A	9 30A	10 35A	10 50A	12 00N	12 55P	1 25P	3 00P	4 00P	5 20P	5 55P	6 20P	8 20P	9 15P
San Diego, CA -Old Town	3													[56] 4 07P	[56] 5 28P		[56] 6 28P		
Solana Beach, CA	26			6 46A	7 39A	8 44A	10 03A	11 08A	11 23A	12 33P	1 28P	1 58P	3 33P	4 34P	5 53P	6 28P	6 53P	8 53P	9 48P
Oceanside, CA (LEGOLAND) [55]	41			7 02A	7 55A	9 00A	10 18A	11 24A	11 39A	12 49P	1 43P	2 13P	3 52P	4 52P	6 08P	6 43P	7 08P	9 08P	10 03P
San Clemente Pier, CA	63												4 13P	5 16P					
San Juan Capistrano, CA	70			7 32A	8 30A	9 30A	10 49A	11 55A	12 10P	1 22P	2 14P	2 47P	4 23P	5 29P	6 39P	7 25P	7 39P	9 42P	10 39P
Laguna Niguel/Mission Viejo, CA	74				8 35A	9 35A			12 15P										
Irvine, CA	83			7 46A	8 45A	9 45A	11 05A	12 09P	12 26P	1 36P	2 28P	3 03P	4 37P	5 41P	6 53P	7 44P	7 53P	9 56P	10 53P
Santa Ana, CA	92		R 5 30A	7 57A	8 56A	9 56A	11 17A	12 20P	12 38P	1 49P	2 39P	3 14P	4 48P	5 52P	7 09P	8 00P	8 09P	10 09P	11 06P
Orange, CA -Metrolink Station	94				9 01A	10 01A													
Anaheim, CA (Disneyland®) [54]	97			8 06A	9 06A	10 06A	11 26A	12 29P	12 47P	1 58P	2 48P	3 23P	4 57P	6 01P	7 18P	8 09P	8 18P	10 18P	11 15P
Fullerton, CA	103		5 50A	8 15A	9 15A	10 15A	11 36A	12 38P	12 57P	2 08P	2 57P	3 34P	5 08P	6 10P	7 27P	8 18P	8 27P	10 27P	11 24P
Los Angeles, CA ✈	128	Ar	[42] 7 15A	8 50A	9 50A	10 50A	12 15P	1 15P	1 35P	2 40P	3 35P	4 05P	5 45P	6 45P	8 05P	8 55P	9 05P	11 05P	11 59P
Los Angeles, CA	128	Dp	7 30A	9 05A			12 30P			2 55P				7 00P		[42] 9 10P	[42] 9 20P		
Glendale, CA	134		7 41A	9 16A			12 42P			3 07P				7 12P		D 9 25P	D 9 40P		
Burbank Airport, CA ✈	142		7 52A	9 28A			12 54P			3 19P				7 24P					
Van Nuys, CA–Amtrak Sta.	147		8 01A	9 37A			1 03P			3 28P				7 33P		D 9 45P	D10 00P		
Chatsworth, CA	157		8 16A	9 50A			1 16P			3 41P				7 46P		D10 05P	D10 20P		
Simi Valley, CA	164		8 45A	10 02A			1 28P			3 59P				7 58P		D10 25P	D10 40P		
Moorpark, CA	175		8 57A	10 15A			1 42P									D10 40P	D10 55P		
Camarillo, CA	186		9 08A	10 26A			1 54P			4 21P				8 27P		D10 50P	D11 05P		
Oxnard, CA	195		9 20A	10 40A			2 06P			4 35P				8 38P		D11 00P	D11 15P		
Ventura, CA	205		9 34A	10 54A			2 19P			4 49P				8 57P		D11 15P	D11 30P		
Carpinteria, CA	221	Ar	9 54A	11 14A			2 45P			5 10P				9 18P		D11 30P	D11 45P		
Santa Barbara, CA	231	Dp	10 11A	D11 33A			D3 04P			5 31P				D 9 38P		11 50P	12 05A		
Santa Barbara, CA	231		10 13A	[43] 11 40A			[43] 3 10P			5 33P				[43] 9 45P					
Goleta, CA	241		10 22A	11 55A			3 20P			5 47P				9 55P					
Solvang, CA	253			12 30P			D4 00P							D10 35P					
Buellton, CA–Burger King				12 25P			D3 55P							D10 30P					
Lompoc, CA–Visitors Center	284						D4 30P												
Lompoc-Surf Sta., CA	300		11 27A							7 02P									
Guadalupe-Santa Maria, CA	326		12 02P				D5 05P			7 39P									
Santa Maria, CA–IHOP	327			1 10P			4 35P							D11 15P					
Grover Beach, CA	338		12 15P	1 35P			D5 30P			7 55P				11 40P					
San Luis Obispo, CA	350		12 45P	2 00P			5 20P			8 30P				12 10A					
San Luis-Cal Poly, CA	1		[45] R 1 00P	2 15P			5 30P			[45] R 8 40P				12 20A					
Atascadero, CA–Denny's	25			2 30P						D 9 00P									
Paso Robles, CA (PT)	35	Ar	1 25P	2 45P			5 55P			9 20P				[64] 12 45A					

Services on Pacific Surfliner® Trains

Coaches: Unreserved.
R Reserved.
B **Pacific Business class Service:** Reserved seat service with complimentary beverages, light snacks and newspaper.
Café: Sandwiches, snacks and beverages on all trains.
Luggage: Checked Baggage service available at select locations; size restriction for carryon luggage is 28" x 22" x 11".
Bicycles: Most Pacific Surfliners are equipped with a limited number of racks—passengers may bring bicycles as unboxed carry-on baggage. Passenger must arrive at station at least 30 minutes before departure and assist with loading and unloading at baggage car. Unboxed bicycles may be put in the bin under connecting Thruway motorcoaches. Amtrak disclaims liability for loss or damage.

Smoking is prohibited entirely on these trains.

○ Unreserved service train.
Thruway and connecting services.

See other side for Symbols and Reference Marks.

The Pacific Surfliner is primarily financed through funds made available by California State Department of Transportation.

For reservations and information, call 1-800-USA-RAIL or your travel agent, or visit Amtrak.com.

Los Angeles • San Diego

Additional late night/early morning Thruway motorcoach service serving Pacific Surfliner route cities. Travel on these buses is reserved and must be part of an itinerary involving a train trip. Advance purchase is required since most stations are unstaffed at the hours the buses operate. Reserved, ticketed customers have priority seating. Passengers traveling north of Los Angeles with valid reservation numbers will be carried; unreserved, ticketed passengers are carried on a space-available basis. Tickets available at Los Angeles, San Diego, Solana Beach and Oceanside 30 minutes before the departure of the buses marked with an asterisk (*) below.

Daily	Daily	Mile	▼		▲	Daily	Daily
		0		Glendale, CA (PT)	Ar		1 45A
		6		Los Angeles, CA	Dp		1 25A
*2 50A	*12 40A	6	Dp	Los Angeles, CA	Ar	*7 15A	*1 15A
D 3 25A	D 1 15A	32		Fullerton, CA		*5 50A	12 30A
D 3 40A	1 30A	42	Ar	Santa Ana, CA		*5 30A	12 10A
D 4 15A		64		San Juan Capistrano, CA			11 35P
D 4 50A		93		Oceanside, CA			*10 55P
D 5 10A		110		Solana Beach, CA			*10 25P
5 35A		135	Ar	San Diego, CA (PT)	Dp		*10 00P

NRPC Form W31–150M–1/21/08 Stock #02-3317J

Schedules subject to change without notice.

AMTRAK

PACIFIC SURFLINER® SOUTHBOUND
SAN LUIS OBISPO–LOS ANGELES–SAN DIEGO

Effective JANUARY 21, 2008

Paso Robles • San Luis Obispo • Santa Barbara • Los Angeles • Orange County • San Diego

Train Number ▶	Mile		562	564	566	768	572	774	578	580	582	784	590	798	592	796
Normal Days of Operation ▶			Mo-Fr	Daily	Daily	Daily	Daily	Daily	Daily	SaSu	Daily	Daily	FrSaSu	Daily	Daily	Daily
Paso Robles, CA (PT)	0	Dp				[64] R 3 15A						[64] R *10 40A		[64] R 12 50P		[64] R 2 40P
Atascadero, CA–Denny's	10															R 3 00P
San Luis-Cal Poly, CA	34					R D3 50A						R *10 10A		1 25P		3 25P
San Luis Obispo, CA	0					4 00A		6 45A				R10 35A		[45] 2 00P		3 50P
Grover Beach, CA	12							7 05A				11 00A		2 20P		4 15P
Santa Maria, CA–IHOP	24					4 35A						R11 25A				4 40P
Guadalupe-Santa Maria, CA	25							7 21A						2 36P		
Lompoc-Surf Sta., CA	51							7 55A						3 10P		
Lompoc, CA–Visitors Center	67											R12 10P				
Buellton, CA–Burger King						5 05A						R12 30P				5 10P
Solvang, CA	98					5 15A						R12 40P				5 20P
Goleta, CA	110					6 30A		9 03A				1 45P		4 15P		6 45P
Santa Barbara, CA	119	Ar				[43] 6 30A						[43] 1 45P				[43] 6 40P
Santa Barbara, CA	119	Dp				6 45A		9 21A			R 12 30P	1 59P		4 29P		6 58P
Carpinteria, CA	129					7 01A		9 37A				2 15P		4 45P		7 14P
Ventura, CA	145					7 22A		9 58A				2 40P		5 06P		7 35P
Oxnard, CA	155				[44]	7 37A		10 12A			1 15P	2 57P		5 20P		7 49P
Camarillo, CA	165				[44]	7 47A		10 23A				3 08P				8 00P
Moorpark, CA	175			[44]	[44]	8 06A		10 34A				3 21P		6 15P		
Simi Valley, CA	186			[44]	[44]	8 20A		10 51A				3 36P		6 29P		8 35P
Chatsworth, CA	194			[44]	[44]	8 35A		11 03A				3 49P		6 44P		8 47P
Van Nuys, CA–Amtrak Sta.	203			[44]	[44]	8 50A	R 9 50A	11 18A			2 35P	4 15P		6 57P		9 04P
Burbank Airport, CA ✈	209			[44]	[44]	8 59A		11 27A				4 25P		7 06P		9 13P
Glendale, CA	216			[44]	[44]	9 10A	10 20A	11 39A				4 37P		7 18P		9 25P
Los Angeles, CA ✈	222	Ar				9 25A	[42] 10 45A	12 10P			[42] 3 35P	4 55P		7 40P		9 45P
Los Angeles, CA	222	Dp	6 05A	7 20A	8 30A	9 40A	11 10A	12 25P	2 00P	3 00P	4 10P	5 10P	7 00P		8 20P	10 10P
Fullerton, CA	248		6 37A	7 52A	9 02A	10 12A	11 42A	12 57P	2 32P	3 32P	4 42P	5 42P	7 32P		8 51P	10 42P
Anaheim, CA (Disneyland®) [54]	253		6 46A	8 01A	9 11A	10 23A	11 51A	1 06P	2 41P	3 41P	4 51P	5 51P	7 41P		9 01P	10 51P
Orange, CA -Metrolink Station	255				9 15A				2 45P							
Santa Ana, CA	258		6 55A	8 10A	9 21A	10 32A	12 00N	1 15P	2 51P	3 50P	5 00P	6 00P	7 50P		9 10P	11 00P
Irvine, CA	268		7 06A	8 24A	9 32A	10 43A	12 13P	1 26P	3 02P	4 04P	5 11P	6 13P	8 04P		9 21P	11 11P
Laguna Niguel/Mission Viejo, CA	277				9 41A				3 11P							
San Juan Capistrano, CA	280		7 20A	8 42A	9 48A	11 02A	12 27P	1 40P	3 17P	4 18P	5 25P	6 27P	8 18P		9 37P	11 25P
San Clemente Pier, CA	288				9 59A	11 12A										
Oceanside, CA (LEGOLAND) [55]	309		7 57A	9 14A	10 24A	11 34A	1 00P	2 13P	3 48P	4 53P	5 56P	6 58P	8 49P		10 08P	11 56P
Solana Beach, CA	325		8 15A	9 29A	10 39A	11 49A	1 19P	2 28P	4 08P	5 10P	6 18P	7 13P	9 04P		10 23P	12 11A
San Diego, CA -Old Town	347			[56] 9 56A	11 06A	[56] 12 16P										12 38A
San Diego, CA (Tijuana) (PT)	350	Ar	8 55A	10 10A	11 20A	12 25P	1 55P	3 10P	4 50P	5 45P	7 00P	7 50P	9 40P		11 05P	12 50A

Symbols and Reference Marks

✈ - Walking distance of main terminal Burbank-Bob Hope Airport. Free shuttle also available. Visit bobhopeairport.com.
- Convenient FlyAway Bus Service Los Angeles Union Station - Los Angeles International Airport. Visit lawa.org.

D Stops only to discharge passengers; train may leave before time shown.
R Stops only to receive passengers.
T Quik-Trak self-serve ticketing kiosk available for credit/debit card sales.

[42] Connection between Thruway motorcoach and train at Los Angeles.
[43] Connection between Thruway motorcoach and train at Santa Barbara.
[44] Metrolink commuter train connection available. Separate ticket required. Call Metrolink at (800) 371-LINK for exact departure times.
[45] Connection between train and Thruway motorcoach at San Luis Obispo station.
[54] Disneyland® is located 2 1/2 miles from Anaheim station.
[55] LEGOLAND is located 8 miles from Oceanside station. Transfers may be made by taxi at passenger's expense.
[56] Will stop at Old Town on Saturdays and Sundays.
[64] Connecting Bus service to/from Oakland - San Luis Obispo - Santa Barbara.
* Connecting services handled by two bus departures.

○ Unreserved service train.
Thruway and connecting services.

See other side for Services on the Pacific Surfliner.

The Pacific Surfliner is primarily financed through funds made available by the California State Department of Transportation.

For reservations and information, call 1-800-USA-RAIL or your travel agent, or visit Amtrak.com.

Schedules subject to change without notice. Amtrak is a registered service mark of the National Railroad Passenger Corp.

NRPC Form W31–150M–1/21/08 Stock #02-3317J

Amtrak Surfliner timetable. Jan, 2008.

The Rail 2 Rail ticketing has made traveling by rail even more convenient for the commuters on both the Metrolink and the Coaster services. Amtrak No. 769 is shown here with the new bi-level equipment as it rolls through Rose Canyon near Miramar Hill on this April 2003 spring day. *Steve Glischinski*

Train No. 454 is shown here at San Diego in April 2003. Note the light-rail train to the left of the photo. California has truly launched the new formula for travel with its coordination between Amtrak and the Commuter Routes as well as the Light Rail lines. *Steve Glischinski*

The Amtrak California trains now provide more service than ever in the history of the Los Angeles–San Diego route. Train No. 578 is rolling across the bridge near Cardiff. *Steve Glischinski*

The new Amtrak California bi-level equipment operates in the push-pull mode, as can be observed with this set of equipment at the Union Station in Los Angeles. This Cab Car also has a baggage compartment and is therefore listed Coach Baggage. The Surfliner trains are equipped with basically the same type of bi-level cars one can observe on the Capitol Corridor as well as the San Joaquins. The set of equipment illustrated here (August 2005) is between runs at the Los Angeles Union Station. *Steve Glischinski*

Side view of Cab Coach Baggage No. 6906. *Erik Frodsham*

Side view of Coach 6453. *Patrick C. Dorin*

Bi-level Coach No. 6404, which was assigned to the Surfliners when its portrait was taken in April 2006. *Erik Frodsham*

Still another type of car on the Surfliners is the Pacific Business Class Car, such as the 6803 illustrated here. *Erik Frodsham*

We begin the Caltrain illustrations with a photo dating back to April 2, 1985, with the GP9 No. 3187, in Caltrain colors, heading up a three-car commuter train in the Southern Pacific colors. Truly a train depicting the transition era from the SP to Caltrain. *Erik Frodsham*

Chapter 5

CALTRAIN: SAN FRANCISCO

Caltrain, the commuter rail service between San Francisco and San Jose, with commute-hour service to Gilroy, took over operation of the line from the SP in 1992. The SP was the only full commuter rail service in the western United States, with some additional operations through subsidiaries such as the Pacific Electric in the Los Angeles area, which was a different type of service. Now service over the former SP line is better then ever under the local operation by Caltrain.

The Peninsula Corridor Joint Powers Board has the responsibility for the Caltrain services on the San Francisco Peninsula and south into Santa Clara County. Caltrain operates the system that has been an important part of passenger transportation for over 144 years as of 2007. The route covers 77 miles from San Francisco to Gilroy and serves 31 stations.

The SP was having many financial difficulties with its commuter rail operations as were most rail lines in North America. To solve the problems, the California Department of Transportation (CalTrans) and the SP made a purchase-of-service agreement. This began in 1980 with the SP operating the trains with subsidies from both the state and the local agency. Caltrans provided the administration, marketing, scheduling and other components for the train services. New developments were soon underway.

The San Francisco Metropolitan Transportation Authority, the San Mateo County Transit District and the Santa Clara Valley Transportation Authority were part of a study that recommended changing the responsibility from the state to the local levels in 1987. The change took place in October 1991 with the creation of the Peninsula Corridor Joint Powers Board. By December, the JPB purchased the SP route between San Francisco and San Jose, and obtained perpetual trackage rights between San Jose and Gilroy.

Caltrain No. 3187 is illustrated here sharing the engine facilities with SP power at San Jose in April 1985. *Erik Frodsham*

In July 1992, the JPB assumed operation of the line from the state and began Caltrain passenger rail operations with a contract with Amtrak to operate the service. This contract continues with Amtrak until June 30, 2009, with two possible one-year extensions.

Passenger traffic for the most part has continued to grow and more trains were added throughout the years. To give one an idea of the growth since 1993, the ridership for the fiscal year in 1993 was 6,875,000. In 2001, it reached 10,510,000. There was a low point in 2004, but ridership growth resumed with over 10,900,000 in fiscal year 2007.

Looking at the ridership per weekday also illustrates the tremendous growth in passenger traffic. In 1969, the Southern Pacific was handling 11,000 passengers per weekday. Moving into the 21st Century, in the year 2002, the average weekday ridership was 30,961. As of 2006, the ridership was 32,291 per weekday and up to 33,841 in February 2007.

The train scheduling is now 96 passenger trains each weekday. The route has been substantially rebuilt by expanding the double track and new stretches of a third main line track near Brisbane and Sunnyvale. All or most of the tracks are signaled for operation in both directions adding new levels of train traffic capacity with a new Centralized Traffic Control System. This has been particularly beneficial for the development of the new train service known as Baby Bullets. These trains are now making the fastest runs between San Jose and San Francisco.

The Baby Bullet Express trains began operation in 2004. The trains were (and are) equipped with new Bombardier Bi-Level Coaches with new MPI MP36PH-3C locomotives. The new Baby Bullet service reduced travel time by stopping at only five stations between San Francisco (4th and King) and San Jose (the Diridon Station). The travel time between San Francisco and San Jose for the express service is 57 minutes compared with 90 minutes for the local

Moving forward into 1988, train No. 65 with new passenger equipment in the Caltrain coloring and lettering is pausing at Millbrae while en route to San Francisco. The stainless steel equipment is part of the fleet as we move through the first decade of the 21ˢᵗ Century. *W. C. Whittaker*

train services. The frequency of the Baby Bullets was increased with two additional trains in May 2005, and another ten in August 2005.

However, the new train service was and is only part of the future service improvements. Plans, either on the drawing board or underway as this is being written in 2007, include several projects. One is to extend the line from the present station in San Francisco with a tunnel 1.3 miles long to the Transbay Terminal providing passengers with easier access to the center of the city. This project involves changing from diesel power to electrification, in part because of the tunnel that is needed to extend the route further downtown.

There is also the plan to reconstruct the Dumbarton rail bridge spanning the southern end of San Francisco Bay. This would provide a new service between the Peninsula and Alameda County in the East Bay. The project would add four new stations to the system: Union City, Fremont-Centerville, Newark, and Menlo Park/East Palo Alto. . This new branch line is tentatively scheduled to be completed in 2012 with six trains in each direction, three to and from San Francisco and the other three to and from San Jose.

The former Southern Pacific commuter route from San Francisco now has more passenger trains than ever in its history. Passengers have several options of train service scheduled throughout the entire day. Train services begin around 5:00 a.m. and extend until after midnight. Weekend services are hourly starting at 8:00 a.m. and continue to midnight on Saturday, and until 9:00 p.m. on Sunday. The services provide a convenient schedule for shopping, entertainment, working on weekends, concerts and athletic games and a whole host of other activities.

The Caltrain System is the way to go.

www.caltrain.com

MOTIVE POWER and PASSENGER EQUIPMENT

The motive power is quite different in 2006 than it was in the late 1960s with the Southern Pacific diesel power. In fact, one could say that there is a far greater number of locomotives in service in 2006 than there was 30 years ago in 1976. Following is a Caltrain Motive Power Roster as of the year 2006

Type	Number Series	Builder	Remarks
F40PH-2	902, 903, 907 910, 914	EMD	Service from 1985 to Present Overhauled by Alstom in 1999.
F40PH-2CAT	900, 901, 904-906 908, 909, 911-913 915-919	EMD	Service from 1985 to Present Originally F40PH-2s, overhauled by Alstom in 1999. Separate HEP generators were installed.
F40PH-2C	920-922	Boise Locomotive Inc.	No. 920 is the Operation Lifesaver Unit.
MP36PH-3C	923-928	Motive Power Inc.	Primarily operated for Baby Bullets
GP9	500, 501	EMD	Service 1980 to 1985 and 2000 to 2004. Used for work train and yard service. SP 3187 received Caltrain's paint scheme.
MP15DC	503, 504	EMD	Service 2003 to Present Used for work train and yard service.

Locomotive Names:

900	San Francisco	901	San Jose	902	San Mateo
903	Santa Clara	904	Palo Alto	905	Sunnyvale
906	Burlingame	907	Mountain View	908	Redwood City
909	Menlo Park	910	Millbrae	911	San Carlos
912	San Bruno	913	Belmont	914	Atherton
915	South San Francisco	916	California	917	Gilroy
918	County of San Mateo	919	County of Santa Clara	920	Morgan Hill
921	San Martin	922	Tamien	925	Jackie Speier

PASSENGER EQUIPMENT ROSTER—2006

Bi-Level Gallery Equipment Built by Nippon Sharyo

Type	Number Series	Number of Cars	Seating Capacity	Year
Gallery Trailers	3800-3825	26	144	1985
	3826-3841	16	148	1985
	3842-3846	10	148	1986
	3847-3851			
Gallery Cab Cars	4000-4020	21	107	1985
	4021	1	88	1999
	4022-4026	5	88	2000
Gallery Trailers	3852-3865	14	122	2000

BOMBARDIER BI-LEVEL CARS

Type	Number Series	Number of Cars	Seating Capacity	Year
Cab	112 to 116	5	124	2002
	117, 118	2	140	2002
Trailers	219 to 230	12	144	2002

Operated primarily on the Baby Bullet Trains

Boise Budd Single Level Cars—Formerly Virginia Railway Express equipment, now out of service on Caltrain and sold to the Grand Canyon Railway

Type	Number Series	Number of Cars	Seating Capacity	Year
Cab Control	800, 803	2	90	1982
Trailers	301, 304-306	12	99	1982
	308, 309, 311			
	314, 401-404			

The following photos and other materials illustrate the Caltrain Commuter Rail Services during the time period of the 1980s to the first decade of the 21st Century—2007.

STATION INFORMATION (Caltrain)

The SAVINGS PER YEAR table illustrates the advantages of taking the commuter train as opposed to driving, not to mention the stress of driving and the loss of productivity at work and the potential of full relaxation at home.

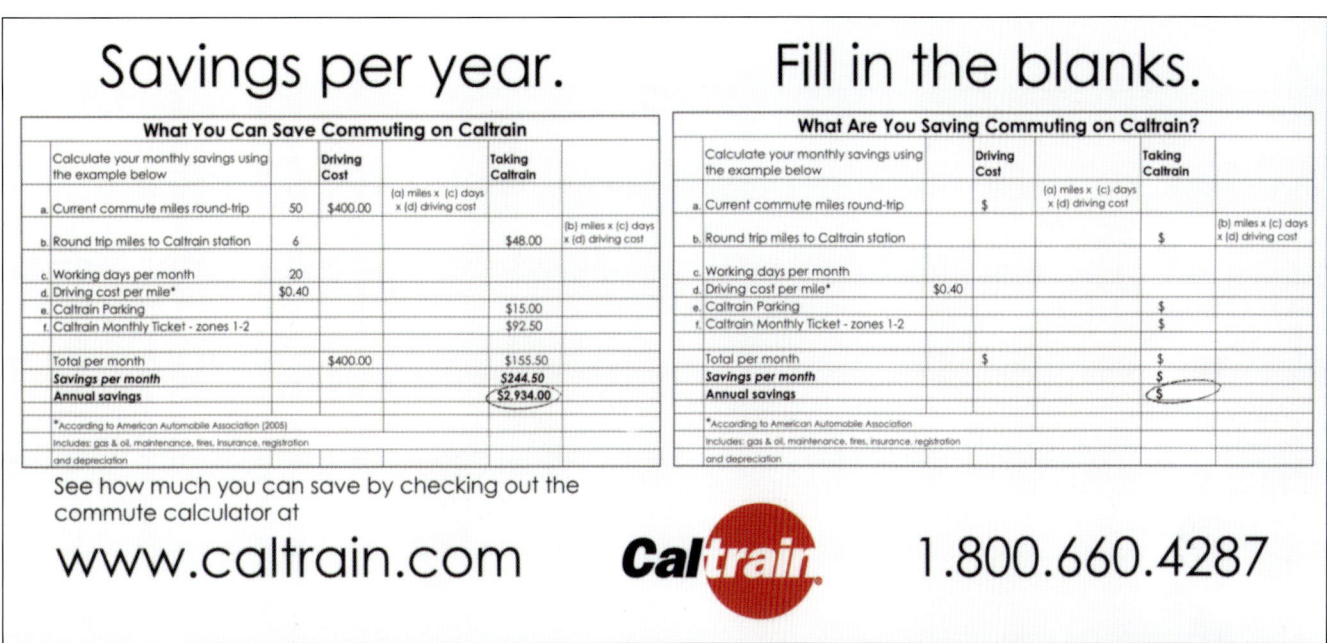

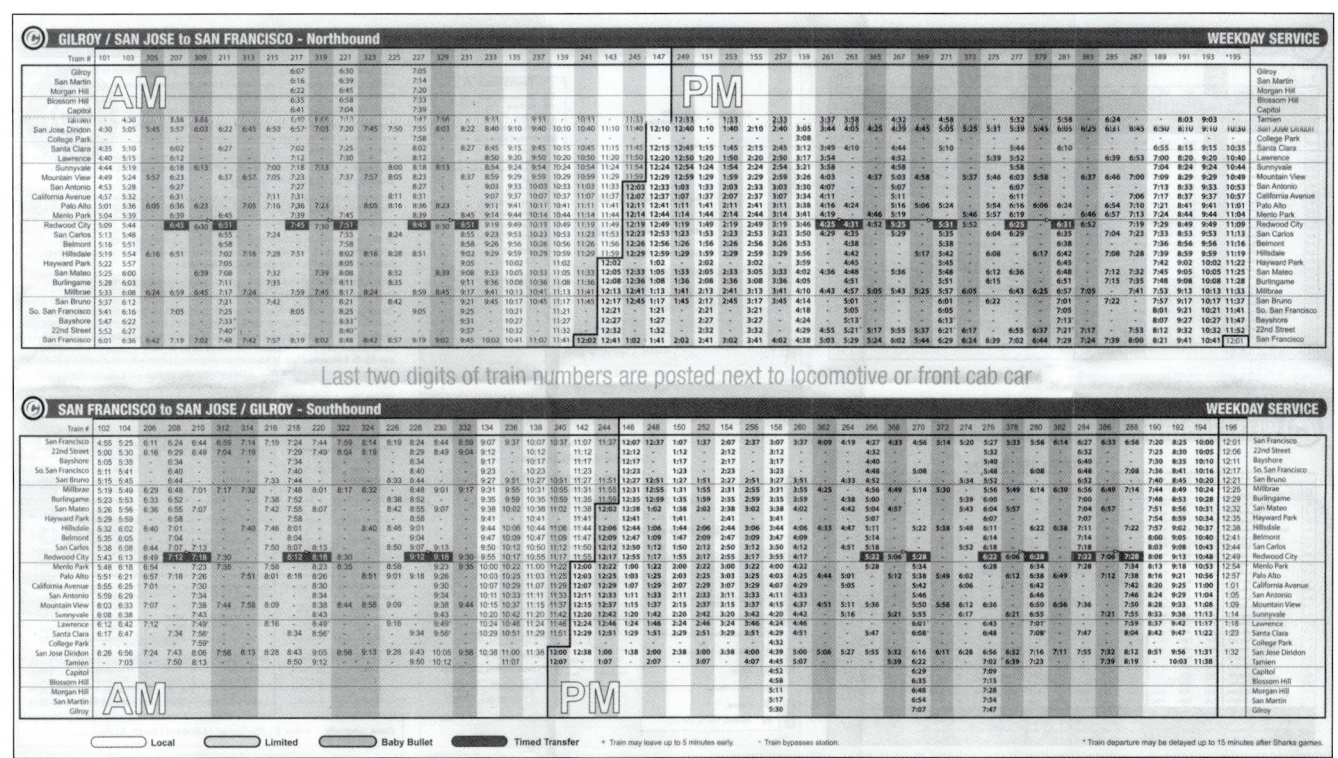

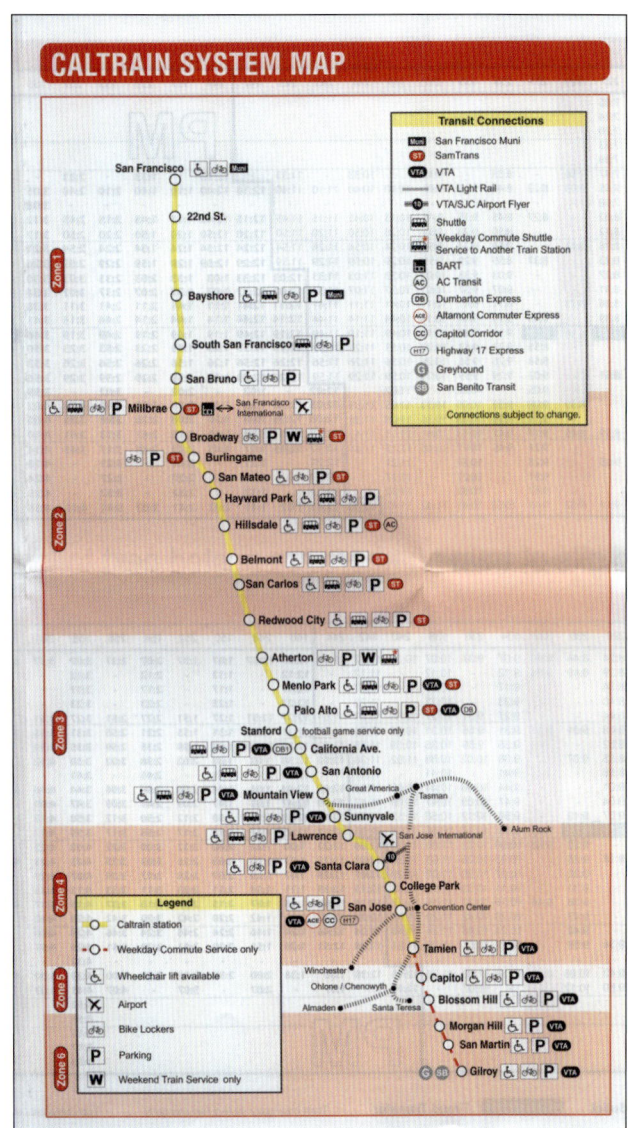

CALTRAIN SYSTEM MAP

Transit Connections

Muni	San Francisco Muni
ST	SamTrans
VTA	VTA
	VTA Light Rail
	VTA/SJC Airport Flyer
	Shuttle
	Weekday Commute Shuttle Service to Another Train Station
BART	BART
AC	AC Transit
DB	Dumbarton Express
ACE	Altamont Commuter Express
CC	Capitol Corridor
H17	Highway 17 Express
G	Greyhound
SB	San Benito Transit

Connections subject to change.

Legend

O	Caltrain station
• - O	Weekday Commute Service only
🦽	Wheelchair lift available
X	Airport
🚲	Bike Lockers
P	Parking
W	Weekend Train Service only

Zone 1 / Zone 2 / Zone 3 / Zone 4 / Zone 5 / Zone 6

San Francisco
22nd St.
Bayshore
South San Francisco
San Bruno
Millbrae — San Francisco International
Broadway
Burlingame
San Mateo
Hayward Park
Hillsdale
Belmont
San Carlos
Redwood City
Atherton
Menlo Park
Palo Alto
Stanford (football game service only)
California Ave.
San Antonio
Mountain View
Sunnyvale
Lawrence
Santa Clara
College Park
San Jose
Tamien
Capitol
Blossom Hill
Morgan Hill
San Martin
Gilroy

Caltrain.

Regional Rail Link

information guide

San Francisco ⟷ San Jose / Gilroy

Caltrain consists include the bi-level gallery, stainless steel cars that replaced the smooth-side equipment from the Southern Pacific. This southbound train is racing over the main line south of San Francisco and is en route to San Jose. *Caltrain Photo*

Caltrain has a number of variations with train consists, such as the new equipment from Bombardier. This photo from April 2007 at San Jose illustrates a train to the left, which has a consist of the gallery cars, while the train to the right is being prepared for departure to San Francisco with the new Bombardier bi-level cars. The colors of the new bi-level coaches include red colors along with the silver scheme. *Patrick C. Dorin*

Train No. 240 from San Francisco has just arrived at Tamien right on time at 12:07 p.m. on April 17, 2007. The train was made up of the stainless-steel gallery coaches. It is being readied to operate as train 249 returning to San Francisco. *Patrick C. Dorin*

This view in April 2007 at San Jose shows a train to the right which will operate to San Francisco in the push mode. One can observe a full train consist to the left between runs prior to returning to the northern end of the line. *Patrick C. Dorin*

With trains operating on a frequency of every 30 minutes during the non-rush hour periods, one can observe several trains lined up for departure from the San Francisco Terminal. *Patrick C. Dorin*

All of the station platforms have yellow lines indicating to passengers NOT to step over the line while waiting for a train. Note the non-skid yellow areas next to the edge of the platform, designed to reduce the possibility of slipping in wet weather while getting on or off the trains. This view is at the San Francisco station. *Patrick C. Dorin*

Gilroy is the southern-most terminal for Caltrain as this is being written in 2007. The Gilroy facilities include trackage for the three scheduled rush hour train consists that stay overnight. The trains in this photo consist of four or five cars. The trains depart Gilroy between 6:07 and 7:05 a.m. and depart San Francisco between 3:07 and 5:27 p.m. as per the timetables in 2007. *Caltrain Photo*

This Caltrain photo illustrates a southbound train en route to San Jose with a five-car consist of the gallery cars. This photo is an excellent example of the type of train services provided by Caltrain. The new level of train services, as compared to the frequencies during the final Southern Pacific days, has brought many new levels of benefits for the entire area. Hence only one of the reasons the train service has been expanded and the extension to Gilroy.

The depot platform level is above the main floor of the San Carlos station as can be observed of the station facilities to the left of the photo. *Caltrain Photo*

Train No. 236 is a "Limited" train meaning it does not stop at all stations between San Francisco and Tamien. The schedule provides a mid-morning service from the Bay Area, and the train is shown here arriving at Santa Clara. *Caltrain Photo*

It is train time at Redwood City with the arrival of a commuter train. Note the attractive station platforms in this photo. *Caltrain Photo*

This view at Redwood City illustrates coach No. 4002, which includes the sign "Bike Car" next to the entry-way. Note the passengers with their bicycles. This photo also illustrates the insignia with the Circle "C" which is now also used by Caltrain. *Caltrain Photo*

Burlingame is only a few minutes south of South San Francisco and has a pleasant and inviting station facility. In this photo passengers are waiting for the southbound train to pull up to a stop and be ready for boarding. *Caltrain Photo*

This view illustrates the front of the Burlingame station with the parking facilities, and the inviting design of the facility. Truly, the station is saying, "Come there is a train for you." *Caltrain Photo*

The Millbrae station is only 24 minutes, train time, south of San Francisco. This station illustrates some of the more traditional depot designs. *Caltrain Photo*

The 918 is heading up a five-car consist at the San Francisco Station in this April 2007 scene. The train will be departing later in the afternoon, but prior to the full rush hour service. *Patrick C. Dorin*

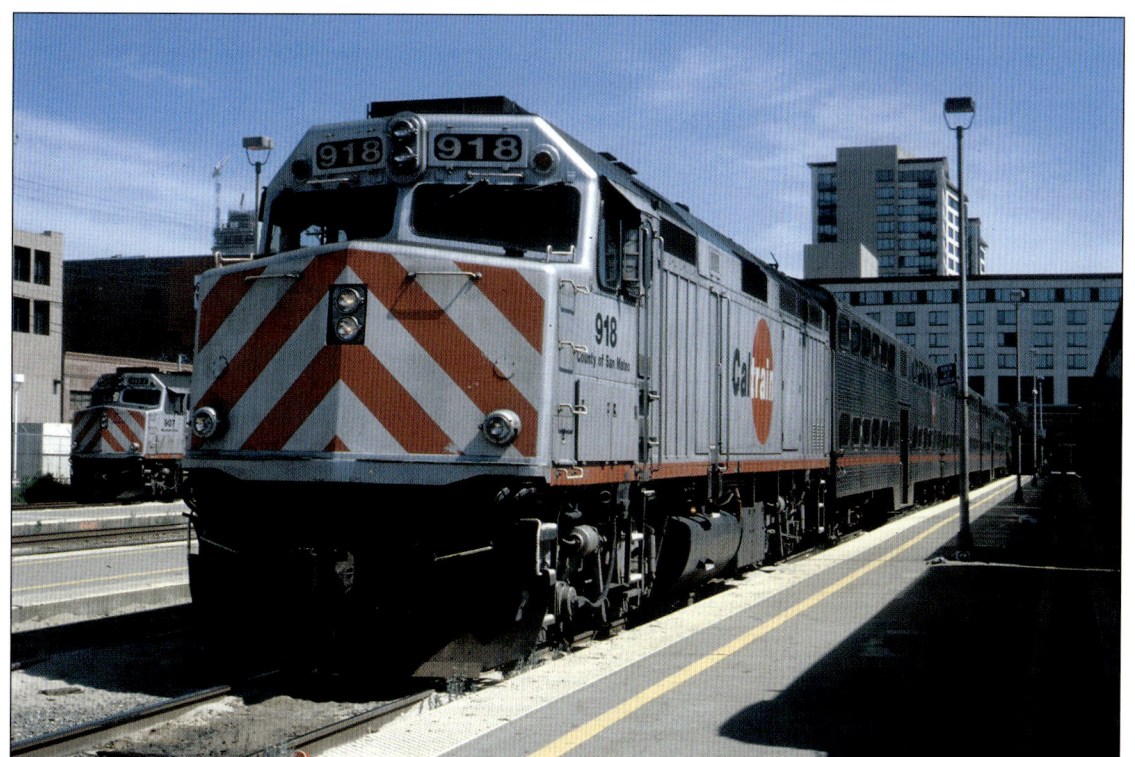

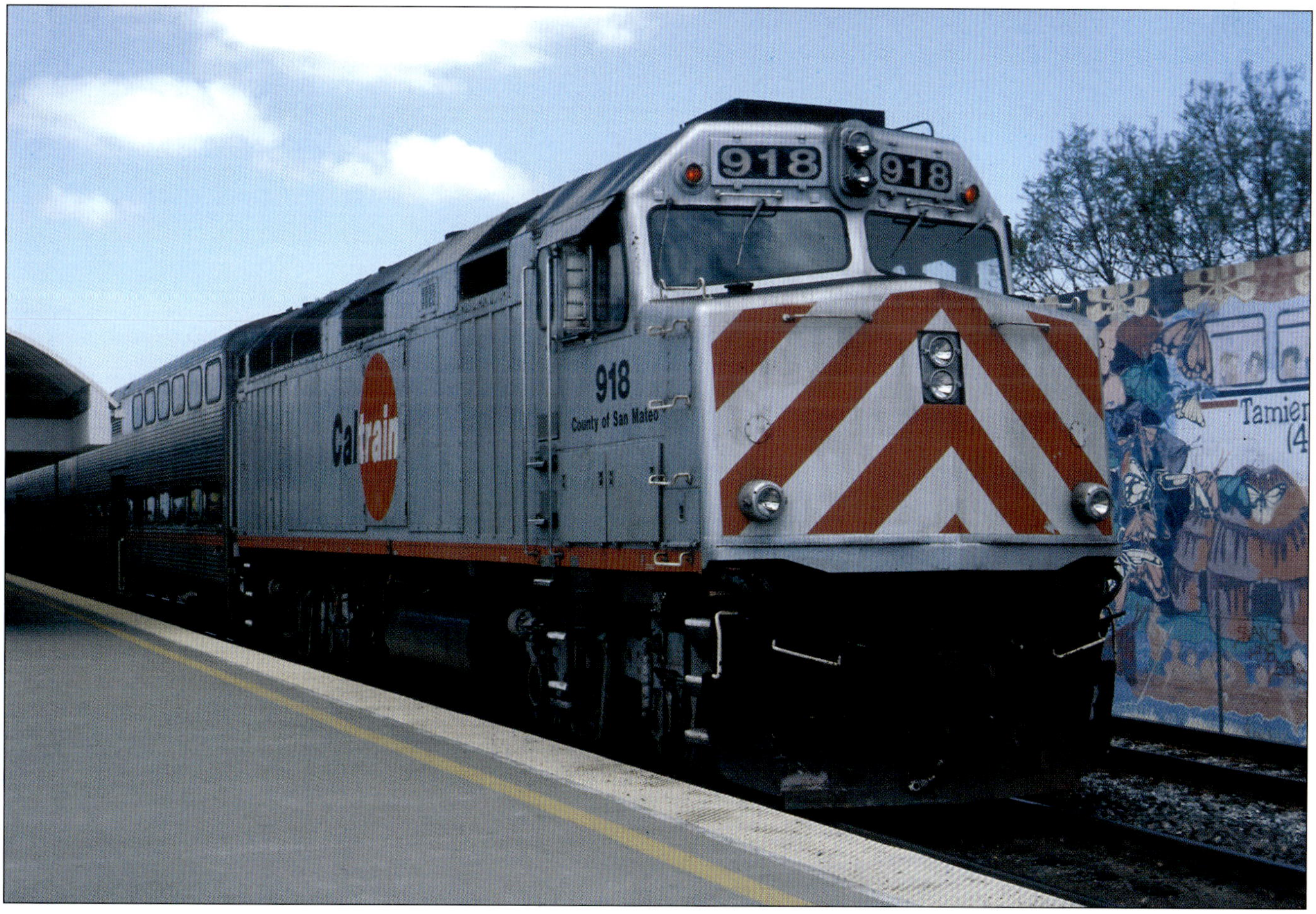

Caltrain No. 918 is a F40PH-2CAT, which was built by Electro-Motive Division in 1987. Here in April 2007, the 918 illustrates the color scheme application in the 21st century, without the stripes but with a large Caltrain insignia. *Patrick C. Dorin*

Caltrain No. 928 is part of the newest fleet of motive power for the commuter train services. It is an MP36PH-3C from Motive Power Industries. The units come in a pleasant color scheme with silver, red and black. In April 2007, No. 928 is heading up a train of Bi-level coaches from Bombardier. *Patrick C. Dorin*

Cab Car No. 116 is shown here on the head-end for a northbound schedule out of San Jose in April 2007. Note how the color scheme matches the colors for the motive power, such as the 928 shown previously in this section. *Patrick C. Dorin*

This view shows the coach seating on the lower level of the Caltrain Gallery coaches. *Patrick C. Dorin*

This view shows the seating arrangement for the upper level, known as the gallery. *Patrick C. Dorin*

It is April 30, 1999, and an eastbound ACE train is at Livermore, California, heading for Stockton. To the far left, one can see a westbound UP freight in the siding as the commuter train rolls by. The motive power is one of the F40PHC-3s, No. 3101, and the train is in the push mode. *Robert Larson*

Chapter 6

THE ALTAMONT COMMUTER EXPRESS: SAN JOSE—STOCKTON

The new Altamont Commuter Express began service on October 19, 1998. The new route connected San Jose on the south with Stockton at the north end, a distance of 85 miles. The new service was the realization of an idea of Robert F. Cabral to create a commuter rail service between the San Joaquin Valley and the Silicon Valley and the South Bay Area.

The new train service began with two rush hour schedules in each direction. The morning trains, No. 1 and 3, departed Stockton at 4:16 and 5:22 a.m. respectively. The trains stopped at Lathrop (Manteca), Tracy, Vasco, Livermore, Pleasanton, Fremont, Great

America and finally San Jose. The westbound train time took 2 hours and 25 minutes.

The evening rush hour schedules, trains 2 and 4, departed San Jose at 4:14 and 5:44 p.m. respectively. The evening running time was also 2 hours and 25 minutes.

The train service was expanded on March 5, 2001, with three trains operating in each direction. The morning rush hour service consisted of trains No. 1 and 3 originating at Stockton with No. 5 originating at Lathrop-Manteca. Trains 1 and 3 made the run to San Jose in 2 hours, 21 minutes. No. 5 took 2 hours,

4 minutes for the schedule from Lathrop-Manteca to San Jose.

The evening service with trains 2 and 4 from San Jose to Stockton made the run in 2 hours, 22 minutes. Train No. 6 made the run from San Jose to Lathrop-Manteca in 2 hours, 8 minutes. The 2001 schedule also included a bus, numbered M 3910, which departed San Jose at 6:45 p.m. and arrived at Stockton at 9:35 p.m. for a 2-hour, 50-minute schedule.

Trains 5 and 6 were eventually expanded to run the full route between San Jose and Stockton. The basic scheduled time as of August 2005 was 2 hours, 10 minutes for five of the six trains. One train, No. 3, took only five minutes more. Basically, the speed and time scheduling for the trains had many improvements from 1998 through 2005. A fourth round-trip train was added on August 28, 2006, as a partnership with Caltrans and Amtrak California. This train not only offers an additional frequency of service for ACE commuters, but it also carries connecting Amtrak Thruway passengers on weekdays between Amtrak's "San Joaquin" trains at Stockton to San Jose and vice versa.

Motive Power and Passenger Equipment

ACE owns and operates six locomotives, type F40PHC-3, number series 3101 to 3106. Three of the units were built in 1997-98 for the start of the new service with the two trains in each direction. Two more units were purchased in the year 2000, numbers 3104 and 3105. The latest unit, No. 3106, was purchased in 2006.

ACE purchased their Bi-level cars from Bombardier in Thunder Bay, Ontario. The initial group ordered for the start up in 1998 was 8 cars, 4 Cab Coaches and 4 Coaches. This has been expanded through the year 2005 with a total fleet of 24 cars.

The roster includes 9 Cab Coaches and 15 Coaches.

All of the equipment contains wheelchair areas and bicycle tiedowns. The seating capacities vary:

Type	Number Series	Built	Remarks
4 Cab Cars	3301—3304	10-1997	137 Seats
4 Coaches	3201—3204	10-1997	136 Seats
	One coach has 122 seats and is known as the Bike Car with 14 seats removed		
5 Cab Cars	3305—3309	9-2000	130 Seats
7 Coaches	3205—3211	9-2000	142 Seats
4 Coaches	3212—3215	4-2005	142 Seats

As of 2007, the maximum number of cars in any given train is six cars. With four trains in each direction, the consists as of late 2006 and into early 2007 were as follows:

Train No.	Number of Cars
1	5
2	3
3	6
4	5
5	5
6	6
7	3
8	5

The annual ridership statistics for the Altamont Commuter Express since 1998 are as follows:

Year	Riders	Operating Days	Average Per Day
1998	67,602	51	1,326
1999	425,116	255	1,667
2000	713,719	254	2,810
2001	923,611	252	3,665
2002	738,969	254	2,909
2003	607,017	253	2,399
2004	644,756	253	2,548
2005	619,873	254	2,440
2006	668,419	252	2,641

The following timetables, maps, charts and photos describe the full operating capabilities and benefits of the new Altamont Commuter Express—a very colorful train. For up-to-date information, the ACE website is *www.acerail.com.*

The cab car No. 3301 is leading the eastbound train with its five-car consist on this neat spring day, April 30th, at Livermore. *Robert Larson*

Stockton is the east end of the ACE route to San Jose. The 3102 and 3101 are shown here at the Stockton facility, while BNSF 1027 East rolls through the area. *Robert Larson*

This view shows the Stockton train facility with the F40PHC-3s numbers 3101 to the right and 3102 to the left. This is how things looked at Stockton in May 1998 when these two photos were taken by Robert Larson.

The color scheme arrangement on the ACE motive power is attractive and provides an invitation for potential passengers. This view illustrates the left side of this particular type of motive power. The 3103's portrait was taken at Stockton in September 1998 by Bryan Griebenow. *William S. Kuba Collection*

This color photo illustrates the application of the three colors to the ACE motive power. No. 3106 is the newest power on the system. *ACE Photo*

An Altamont Commuter Express train is off in the distance with the power windmills in the background as it crosses a bridge with highway traffic flowing underneath. *ACE Photo*

The entryways on the Bombardier bi-level coaches are very convenient for loading and unloading passengers. In this photo, the doors are open and ready for passengers. *ACE Photo*

Passengers are ready to board. *ACE Photo*

This photo shows passengers getting off a train, and depending upon the particular station stop, the folks will either be boarding buses or heading for their cars for the rest of the way to their destination. The coordination with bus services increases the availability of train service for wider geographic areas. *ACE Photo*

The sun has put the ACE bi-level coaches in the spot light, which is one of nature's ways of saying, "Take the train—it will reduce congestion and pollution." *ACE Photo*

The ACE train is moving fast as the red sunset finishes up the day. *ACE Photo*

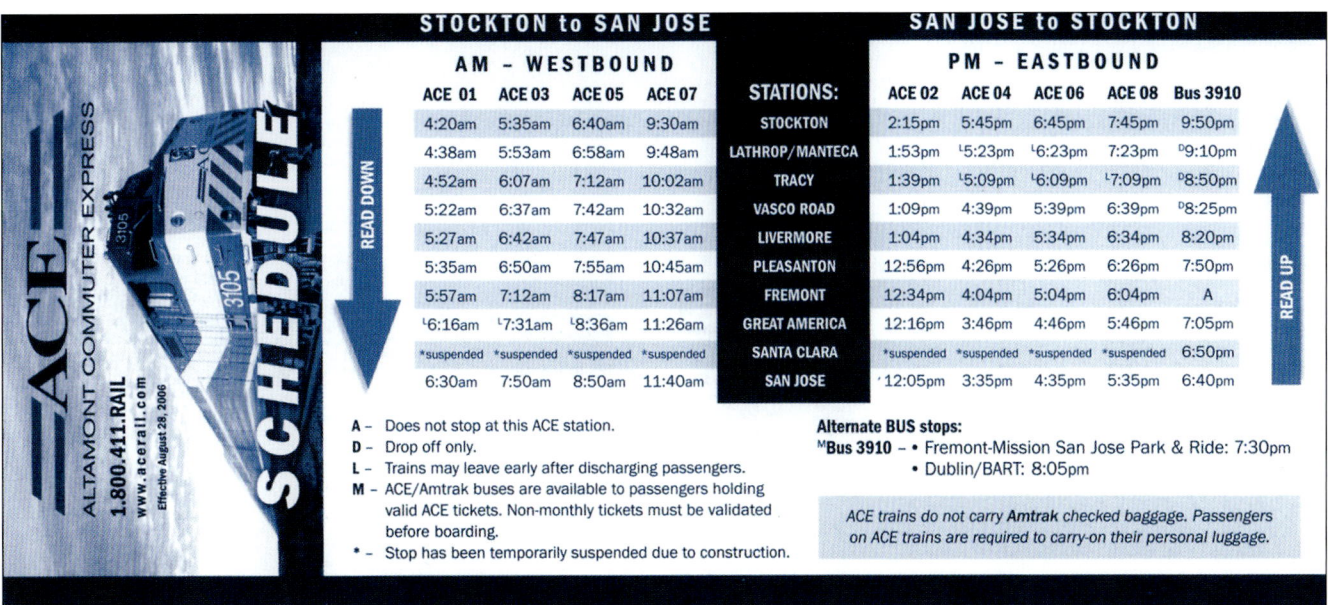

STOCKTON to SAN JOSE — AM – WESTBOUND (READ DOWN)

STATIONS:	ACE 01	ACE 03	ACE 05	ACE 07
STOCKTON	4:20am	5:35am	6:40am	9:30am
LATHROP/MANTECA	4:38am	5:53am	6:58am	9:48am
TRACY	4:52am	6:07am	7:12am	10:02am
VASCO ROAD	5:22am	6:37am	7:42am	10:32am
LIVERMORE	5:27am	6:42am	7:47am	10:37am
PLEASANTON	5:35am	6:50am	7:55am	10:45am
FREMONT	5:57am	7:12am	8:17am	11:07am
GREAT AMERICA	ᴸ6:16am	ᴸ7:31am	ᴸ8:36am	11:26am
SANTA CLARA	*suspended	*suspended	*suspended	*suspended
SAN JOSE	6:30am	7:50am	8:50am	11:40am

SAN JOSE to STOCKTON — PM – EASTBOUND (READ UP)

STATIONS:	ACE 02	ACE 04	ACE 06	ACE 08	Bus 3910
STOCKTON	2:15pm	5:45pm	6:45pm	7:45pm	9:50pm
LATHROP/MANTECA	1:53pm	ᴸ5:23pm	ᴸ6:23pm	7:23pm	ᴰ9:10pm
TRACY	1:39pm	ᴸ5:09pm	ᴸ6:09pm	ᴸ7:09pm	ᴰ8:50pm
VASCO ROAD	1:09pm	4:39pm	5:39pm	6:39pm	ᴰ8:25pm
LIVERMORE	1:04pm	4:34pm	5:34pm	6:34pm	8:20pm
PLEASANTON	12:56pm	4:26pm	5:26pm	6:26pm	7:50pm
FREMONT	12:34pm	4:04pm	5:04pm	6:04pm	A
GREAT AMERICA	12:16pm	3:46pm	4:46pm	5:46pm	7:05pm
SANTA CLARA	*suspended	*suspended	*suspended	*suspended	6:50pm
SAN JOSE	12:05pm	3:35pm	4:35pm	5:35pm	6:40pm

A – Does not stop at this ACE station.
D – Drop off only.
L – Trains may leave early after discharging passengers.
M – ACE/Amtrak buses are available to passengers holding valid ACE tickets. Non-monthly tickets must be validated before boarding.
* – Stop has been temporarily suspended due to construction.

Alternate BUS stops:
ᴹ**Bus 3910** – • Fremont-Mission San Jose Park & Ride: 7:30pm
• Dublin/BART: 8:05pm

*ACE trains do not carry **Amtrak** checked baggage. Passengers on ACE trains are required to carry-on their personal luggage.*

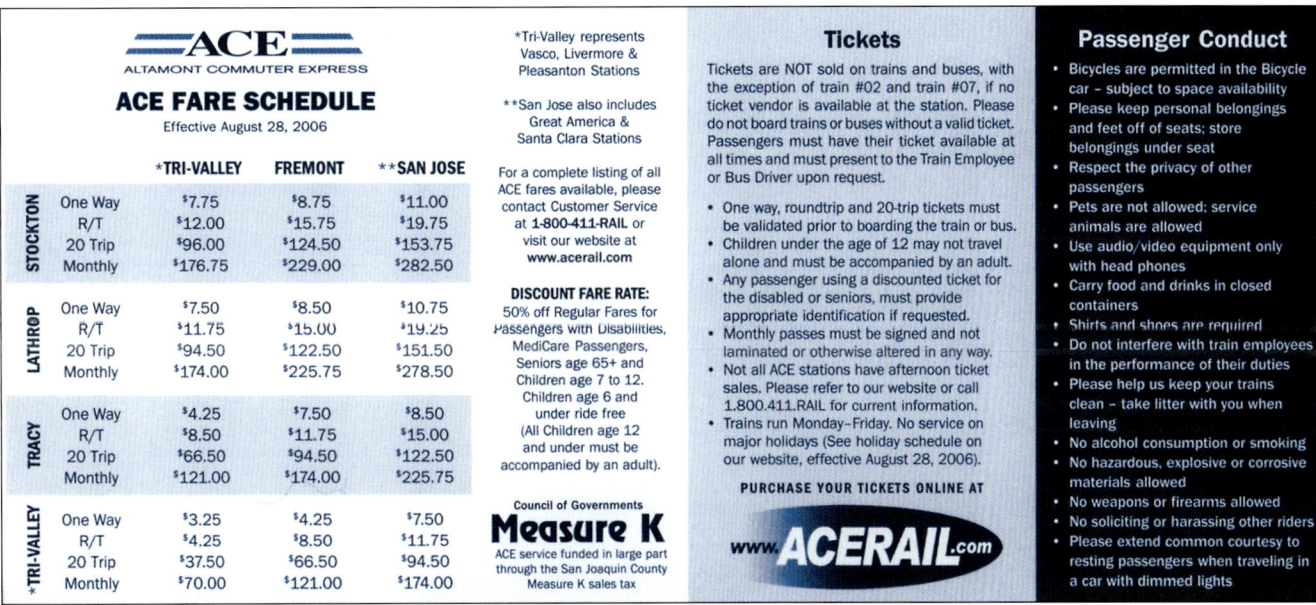

ACE FARE SCHEDULE
Effective August 28, 2006

		*TRI-VALLEY	FREMONT	**SAN JOSE
STOCKTON	One Way	$7.75	$8.75	$11.00
	R/T	$12.00	$15.75	$19.75
	20 Trip	$96.00	$124.50	$153.75
	Monthly	$176.75	$229.00	$282.50
LATHROP	One Way	$7.50	$8.50	$10.75
	R/T	$11.75	$15.00	$19.25
	20 Trip	$94.50	$122.50	$151.50
	Monthly	$174.00	$225.75	$278.50
TRACY	One Way	$4.25	$7.50	$8.50
	R/T	$8.50	$11.75	$15.00
	20 Trip	$66.50	$94.50	$122.50
	Monthly	$121.00	$174.00	$225.75
TRI-VALLEY	One Way	$3.25	$4.25	$7.50
	R/T	$4.25	$8.50	$11.75
	20 Trip	$37.50	$66.50	$94.50
	Monthly	$70.00	$121.00	$174.00

*Tri-Valley represents Vasco, Livermore & Pleasanton Stations

**San Jose also includes Great America & Santa Clara Stations

For a complete listing of all ACE fares available, please contact Customer Service at **1-800-411-RAIL** or visit our website at **www.acerail.com**

DISCOUNT FARE RATE:
50% off Regular Fares for Passengers with Disabilities, MediCare Passengers, Seniors age 65+ and Children age 7 to 12. Children age 6 and under ride free (All Children age 12 and under must be accompanied by an adult).

Council of Governments
Measure K
ACE service funded in large part through the San Joaquin County Measure K sales tax

Tickets

Tickets are NOT sold on trains and buses, with the exception of train #02 and train #07, if no ticket vendor is available at the station. Please do not board trains or buses without a valid ticket. Passengers must have their ticket available at all times and must present to the Train Employee or Bus Driver upon request.

- One way, roundtrip and 20-trip tickets must be validated prior to boarding the train or bus.
- Children under the age of 12 may not travel alone and must be accompanied by an adult.
- Any passenger using a discounted ticket for the disabled or seniors, must provide appropriate identification if requested.
- Monthly passes must be signed and not laminated or otherwise altered in any way.
- Not all ACE stations have afternoon ticket sales. Please refer to our website or call 1.800.411.RAIL for current information.
- Trains run Monday–Friday. No service on major holidays (See holiday schedule on our website, effective August 28, 2006).

PURCHASE YOUR TICKETS ONLINE AT www.ACERAIL.com

Passenger Conduct

- Bicycles are permitted in the Bicycle car – subject to space availability
- Please keep personal belongings and feet off of seats; store belongings under seat
- Respect the privacy of other passengers
- Pets are not allowed; service animals are allowed
- Use audio/video equipment only with head phones
- Carry food and drinks in closed containers
- Shirts and shoes are required
- Do not interfere with train employees in the performance of their duties
- Please help us keep your trains clean – take litter with you when leaving
- No alcohol consumption or smoking
- No hazardous, explosive or corrosive materials allowed
- No weapons or firearms allowed
- No soliciting or harassing other riders
- Please extend common courtesy to resting passengers when traveling in a car with dimmed lights

The August 28, 2006, ACE schedule shows the operating times and station stops for the four trains in each direction. Note the bus schedule, Bus 3910, illustrating a fifth departure from San Jose at the end of the workday.

Although the Capitol Corridor is a basic intercity train operation, the trains do provide a commuter-type service. There are quite a few daily passengers traveling to and from work in Sacramento, the San Francisco Bay Area as well as San Jose. This is the reason that the Capitol Corridor service is part of this book. This photo was taken in the rain April 13, 2007, at Sacramento while a set of equipment laid over between runs to and from Oakland and/or San Jose. *Patrick C. Dorin*

Chapter 7
AMTRAK'S CAPITOL CORRIDOR: SAN JOSE—SACRAMENTO

The Capitol Corridor, on the one hand could be considered an Intercity Train Service, which it is. The route extends from San Jose, California, to Auburn, California; a distance of 172 miles. However, most of the train service is between Oakland and Sacramento, a distance of 90 miles. This 90-mile category is similar to commuter train services on other routes in the United States, one example of which is the Northern Indiana Commuter Transportation District between Chicago and South Bend, Indiana.

It turns out that several people can use the Cor-

ridor service for commuting as well as for traveling to various locations for shopping and entertainment activities. The summer 2007 Amtrak California timetables show three morning eastbound trains arriving in Sacramento in the 6:30 to 8:30 a.m. time period. There are three westbound evening rush hour departures from Sacramento between 4:30 and 6:45 p.m.

Morning westbound arrivals at Oakland include four trains between 6:00 and 9:00 a.m. Evening eastbound departures from Oakland include four trains between 4:00 and 7:00 p.m.

San Jose train services convenient for commuters to and from work include two westbound train arrivals between 7:30 and 8:45 a.m. Eastbound rush hour departures from San Jose include two trains between 4:00 and 6:00 p.m. San Jose also has the advantage of the CalTrain services to San Francisco as well as the Altamont Commuter Express route. San Jose is another city that can lay claim to being served by three rail passenger carriers, similar to Oceanside in Southern California.

There is a well-rounded service level throughout the day on the Capitol Corridor. The summer 2007 timetables shows 16 trains each way on weekdays between Oakland and Sacramento, while there are 7 trains each way between Oakland and San Jose. There is one train in each direction east of Sacramento to and from Auburn.

On weekends there are 11 trains each way between Sacramento and Oakland, and 7 each way between Oakland and San Jose. The weekend service to and from Auburn is the single train in each direction.

The current train frequencies simply dwarf the May 1971 schedule. The Chicago–Oakland train, and the Los Angeles–Seattle trains operated only tri-weekly between Oakland and Davis (to the north) and Sacramento (to the east). However, there was one train in each direction between Oakland and San Jose. These were part of what eventually became the Coast Starlight and the California Zephyr.

Ridership on the Capitol Corridor has grown. For example, daily ridership was averaging over 3,760 passengers for the 151 days between October 2006 and February 2007. This compares with 3,362 per day during the same period the previous year. These figures are based on the ridership counts in the new *Passenger Train Journal*, published in the 2nd Quarter 2007, page 8. (The *Journal* provides ridership statistics and up-to-date developments in rail passenger service. It is an excellent source for new information in North America.)

The Corridor trains are now equipped with a new concept in bi-level passenger equipment built by M-K Amrail Company in the late 1990s. The equipment includes Coaches and Food Service cars offering sandwiches, snacks and beverages. A Quiet Car is part of the consist for an atmosphere free of cell phones and pagers. Public phone service is available on the train. Another interesting aspect about the Corridor equipment is that all trains are equipped to

handle bicycles. Passenger may bring bicycles as unboxed, carry-on baggage. Many commuter services provide space for bicycles.

The following photos and tables illustrate the Capitol Corridor Services with the type of equipment and motive power now providing the State of California with a new level of appropriate transportation.

Readers may wish to refer to the November 1997 issue of *Model Railroader* magazine for diagrams of the Amtrak California equipment.

One of the important advantages of the Capitol Corridor is the co-ordination with bus services. This Amtrak California bus is exchanging passengers with the trains at Emeryville in April 2007. *Patrick C. Dorin*

An angle view of a Bi-level Amtrak California coach, No. 8011. Such equipment operates on all three of the California routes. *Erik Frodsham*

The food service equipment is similar to the coach designs as can be observed with No. 8804 in the Amtrak California color scheme in June 2006. *Erik Frodsham*

An Amtrak California train is laying over at Oakland between runs on the Capitol Corridor in May 1999. The consist is a typical four-car train over the route. *Robert Larson*

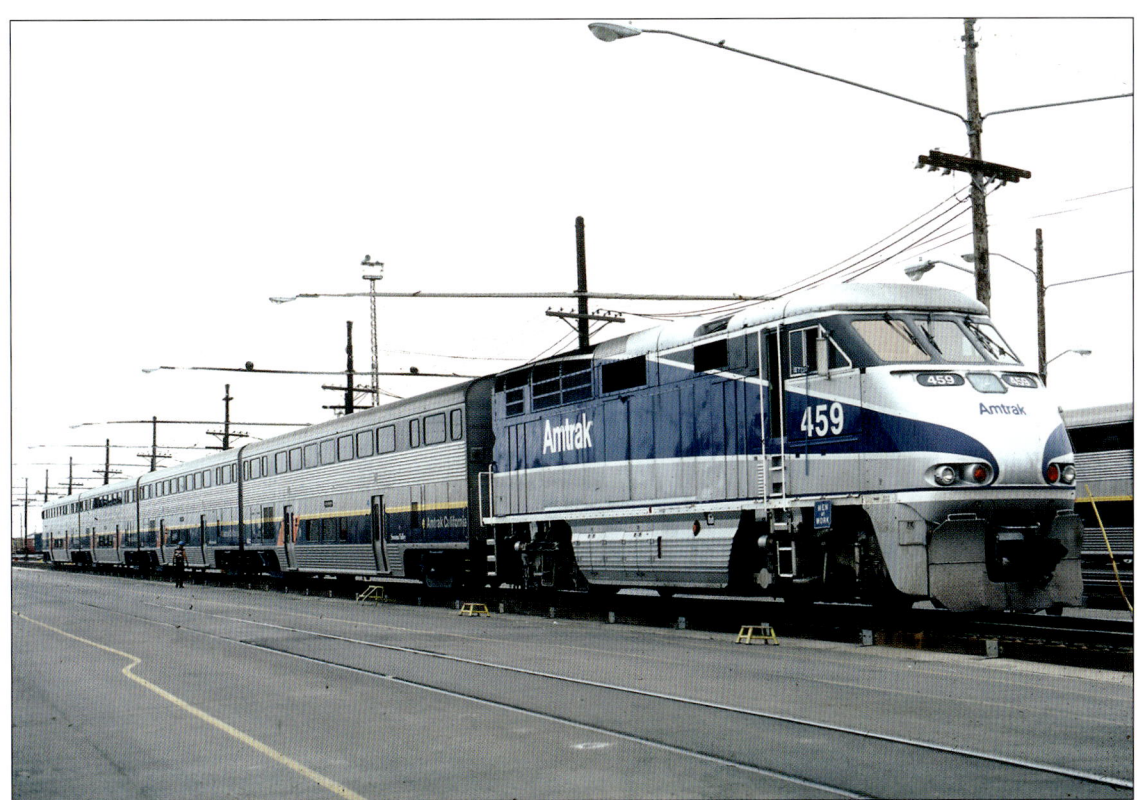

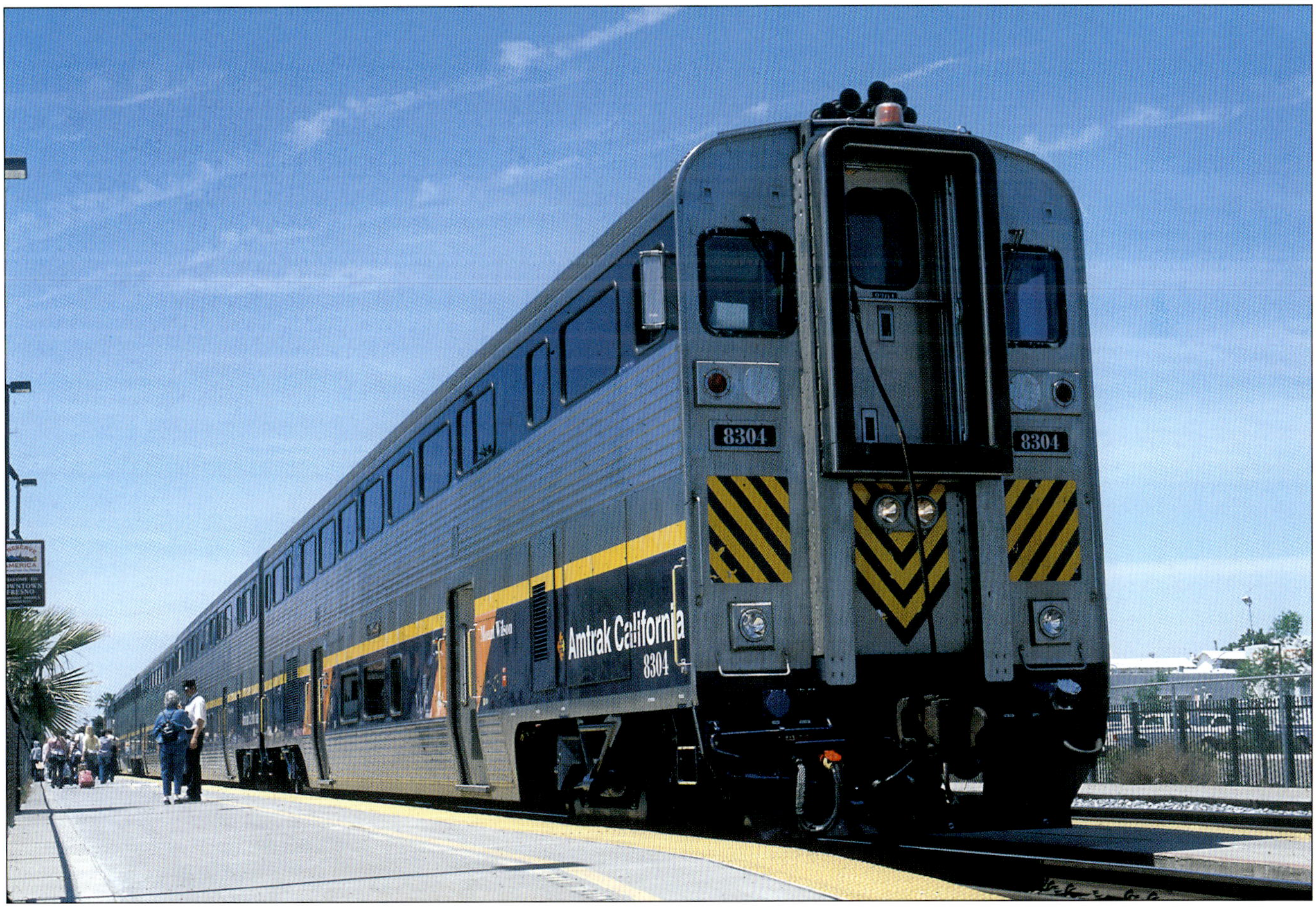

The Capitol Corridor and other California routes operate with the push-pull concept. This photo shows Cab Car 8304 on a San Joaquin at Fresno, California, May 2006. *Erik Frodsham*

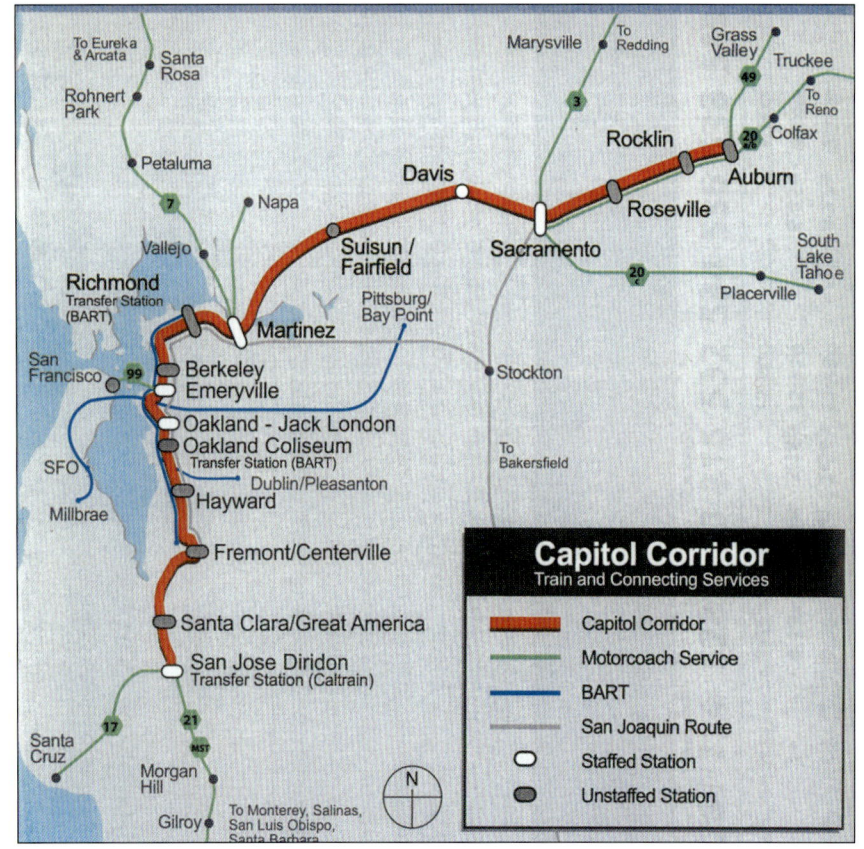

One example of the motive power on the Amtrak California service is No. 2051, a Dash 8-32BW. *William S. Kuba*

Map of the Capitol Corridor. *Amtrak California, May 2007*

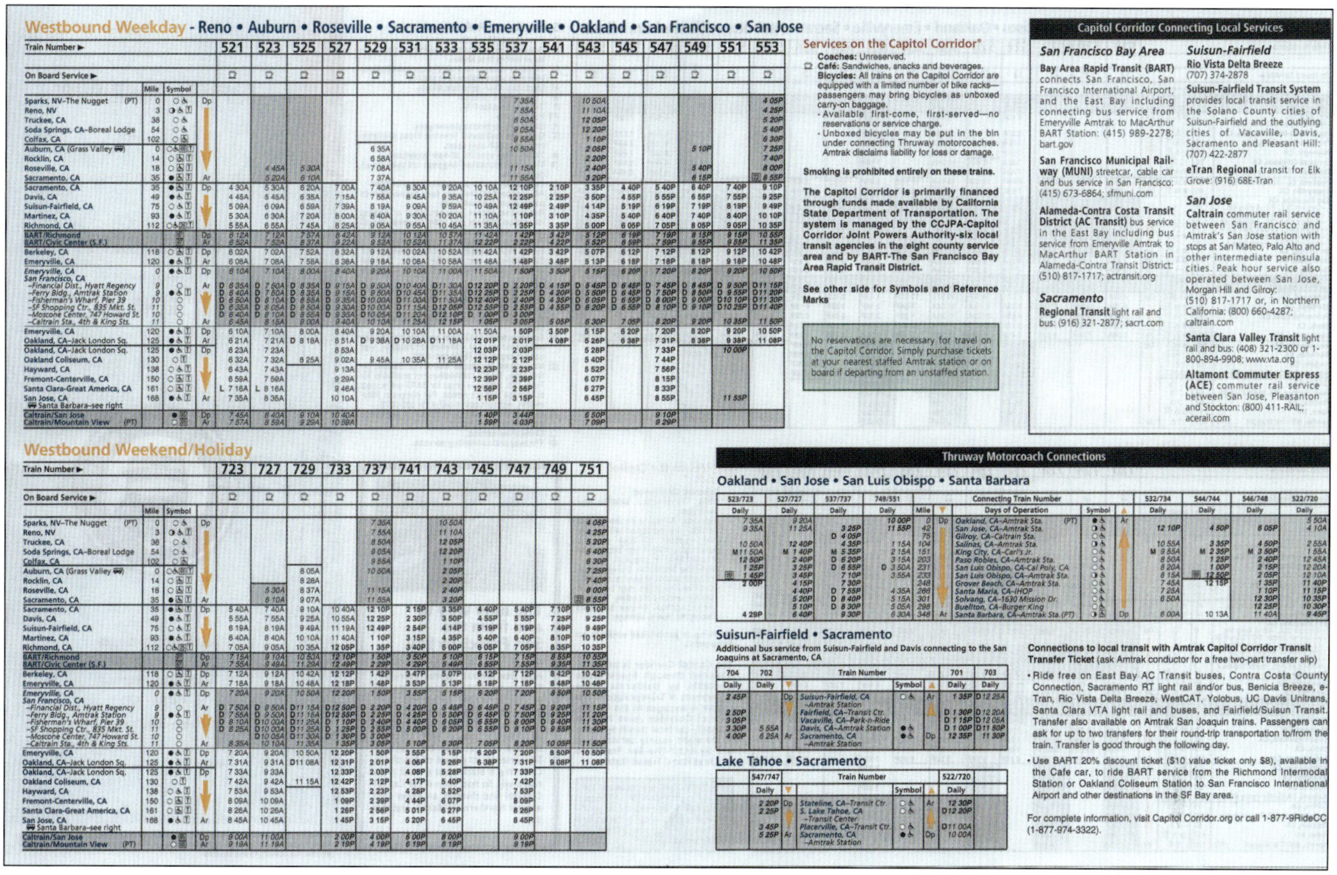

Westbound Weekday - Reno • Auburn • Roseville • Sacramento • Emeryville • Oakland • San Francisco • San Jose

Train Numbers: 521 523 525 527 529 531 533 535 537 541 543 545 547 549 551 553

Stations (selected): Sparks, NV–The Nugget; Reno, NV; Truckee, CA; Soda Springs, CA–Boreal Lodge; Colfax, CA; Auburn, CA (Grass Valley); Rocklin, CA; Roseville, CA; Sacramento, CA; Davis, CA; Suisun-Fairfield, CA; Martinez, CA; Richmond, CA; BART/Richmond; Berkeley, CA; Emeryville, CA; San Francisco, CA (Financial Dist., Hyatt Regency; Ferry Bldg., Amtrak Station; Fisherman's Wharf, Pier 39; SF Shopping Ctr., 835 Mkt. St.; Moscone Center, 747 Howard St.; Caltrain Sta., 4th & King Sts.); Oakland, CA–Jack London Sq.; Oakland Coliseum, CA; Hayward, CA; Fremont-Centerville, CA; Santa Clara-Great America, CA; San Jose, CA; Santa Barbara–see right; Caltrain/San Jose; Caltrain/Mountain View

Westbound Weekend/Holiday

Train Numbers: 723 727 729 733 737 741 743 745 747 749 751

Services on the Capitol Corridor*

Coaches: Unreserved.
Café: Sandwiches, snacks and beverages.
Bicycles: All trains on the Capitol Corridor are equipped with a limited number of bike racks—passengers may bring bicycles as unboxed carry-on baggage.
- Available first-come, first-served—no reservations or service charge.
- Unboxed bicycles may be put in the bin under connecting Thruway motorcoaches. Amtrak disclaims liability for loss or damage.

Smoking is prohibited entirely on these trains.

The Capitol Corridor is primarily financed through funds made available by California State Department of Transportation. The system is managed by the CCJPA-Capitol Corridor Joint Powers Authority-six local transit agencies in the eight county service area and by BART-The San Francisco Bay Area Rapid Transit District.

See other side for Symbols and Reference Marks

No reservations are necessary for travel on the Capitol Corridor. Simply purchase tickets at your nearest staffed Amtrak station or on board if departing from an unstaffed station.

Capitol Corridor Connecting Local Services

San Francisco Bay Area
Bay Area Rapid Transit (BART) connects San Francisco, San Francisco International Airport, and the East Bay including connecting bus service from Emeryville Amtrak to MacArthur BART Station: (415) 989-2278; bart.gov

San Francisco Municipal Railway (MUNI) streetcar, cable car and bus service in San Francisco: (415) 673-6864; sfmuni.com

Alameda-Contra Costa Transit District (AC Transit) bus service in the East Bay including bus service from Emeryville Amtrak to MacArthur BART Station in Alameda-Contra Costa Transit District: (510) 817-1717; actransit.org

Sacramento
Regional Transit light rail and bus: (916) 321-2877; sacrt.com

Suisun-Fairfield
Rio Vista Delta Breeze (707) 374-2878
Suisun-Fairfield Transit System provides local transit service in the Solano County cities of Suisun-Fairfield and the outlying cities of Vacaville, Davis, Sacramento and Pleasant Hill: (707) 422-2877
eTran Regional transit for Elk Grove: (916) 68E-Tran

San Jose
Caltrain commuter rail service between San Francisco and Amtrak's San Jose station with stops at San Mateo, Palo Alto and other intermediate peninsula cities. Peak hour service also operated between San Jose, Morgan Hill and Gilroy: (510) 817-1717 or, in Northern California: (800) 660-4287; caltrain.com

Santa Clara Valley Transit light rail and bus: (408) 321-2300 or 1-800-894-9908; www.vta.org

Altamont Commuter Express (ACE) commuter rail service between San Jose, Pleasanton and Stockton: (800) 411-RAIL; aceraill.com

Thruway Motorcoach Connections
Oakland • San Jose • San Luis Obispo • Santa Barbara

Suisun-Fairfield • Sacramento
Additional bus service from Suisun-Fairfield and Davis connecting to the San Joaquins at Sacramento, CA
Train Numbers: 704 702 / 701 703

Lake Tahoe • Sacramento
Train Numbers: 547/747 / 522/720

Connections to local transit with Amtrak Capitol Corridor Transit Transfer Ticket (ask Amtrak conductor for a free two-part transfer slip)
- Ride free on East Bay AC Transit buses, Contra Costa County Connection, Sacramento RT light rail and/or bus, Benicia Breeze, e-Tran, Rio Vista Delta Breeze, WestCAT, Yolobus, UC Davis Unitrans, Santa Clara VTA light rail and buses, and Fairfield/Suisun Transit. Transfer also available on Amtrak San Joaquin trains. Passengers can ask for up to two transfers for their round-trip transportation to/from the train. Transfer is good through the following day.
- Use BART 20% discount ticket ($10 value ticket only $8), available in the Cafe car, to ride BART service from the Richmond Intermodal Station or Oakland Coliseum Station to San Francisco International Airport and other destinations in the SF Bay area.

For complete information, visit Capitol Corridor.org or call 1-877-9RideCC (1-877-974-3322).

Capitol Corridor Schedules for Weekdays and Weekends. *Amtrak California, May 2007*

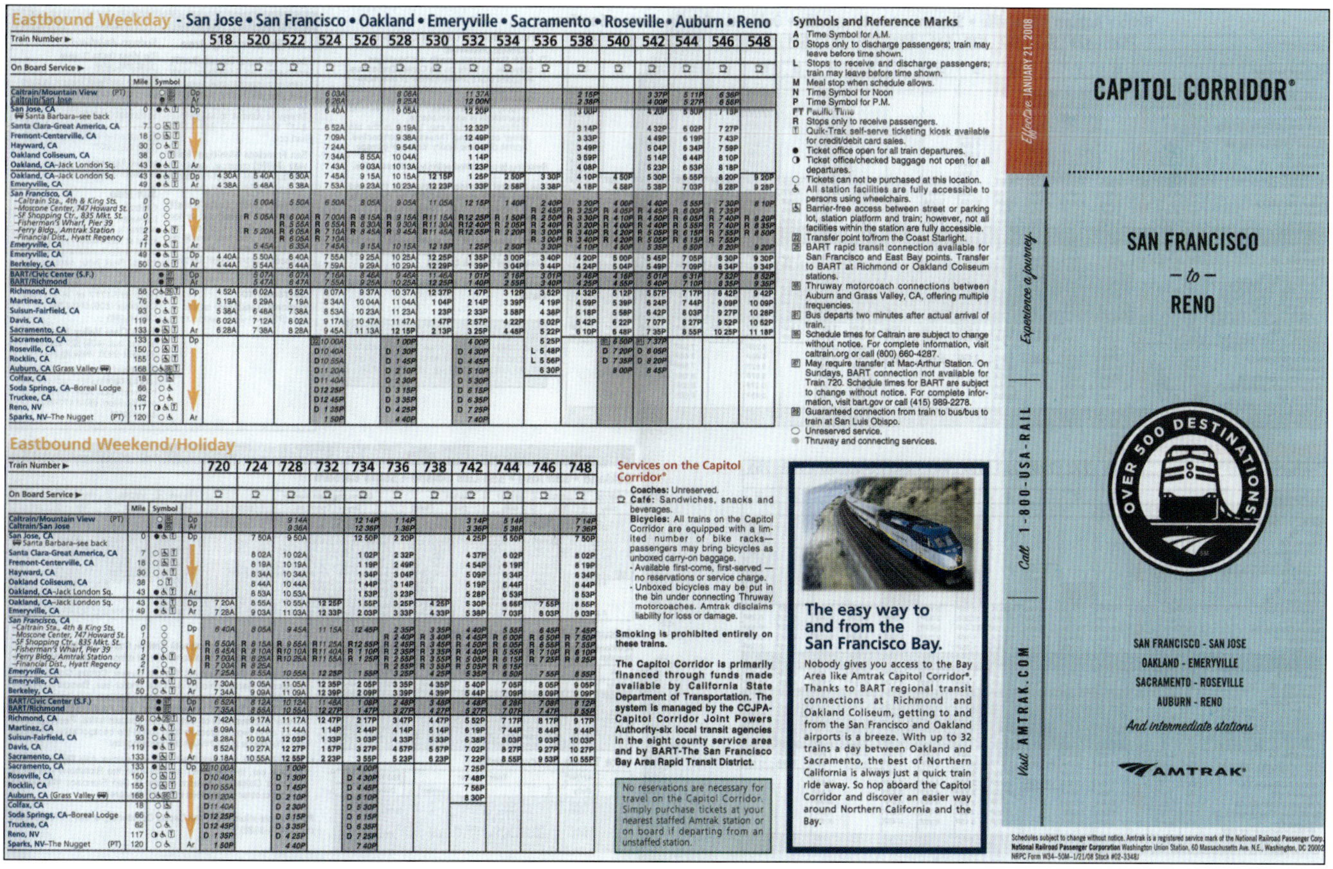

Eastbound Weekday - San Jose • San Francisco • Oakland • Emeryville • Sacramento • Roseville • Auburn • Reno

Train Numbers: 518 520 522 524 526 528 530 532 534 536 538 540 542 544 546 548

Eastbound Weekend/Holiday

Train Numbers: 720 724 728 732 734 736 738 742 744 746 748

Symbols and Reference Marks
A Time Symbol for A.M.
D Stops only to discharge passengers; train may leave before time shown.
L Stops to receive and discharge passengers; train may leave before time shown.
M Meal stop when schedule allows.
N Time Symbol for Noon
P Time Symbol for P.M.
R Flag stop.
T Stops only to receive passengers.
⬜ Quik-Trak self-serve ticketing kiosk available for credit/debit card sales.
● Ticket office open for all train departures.
○ Ticket office/checked baggage not open for all departures.
◇ Tickets can not be purchased at this location.
♿ All station facilities are fully accessible to persons using wheelchairs.
Ⓑ Barrier-free access between street or parking lot, station platform and train; however, not all facilities within the station are fully accessible.
Ⓣ Transfer point to/from the Coast Starlight.
Ⓑ BART rapid transit connection available for San Francisco and East Bay points. Transfer to BART at Richmond or Oakland Coliseum stations.
Ⓣ Thruway motorcoach connections between Auburn and Grass Valley, CA, offering multiple frequencies.
Ⓑ Bus departs two minutes after actual arrival of train.
Ⓒ Schedule times for Caltrain are subject to change without notice. For complete information, visit caltrain.org or call (800) 660-4287.
Ⓜ May require transfer at Mac-Arthur Station. On Sundays, BART connection not available for Train 720. Schedule times for BART are subject to change without notice. For complete information, visit bart.gov or call (415) 989-2278.
○ Guaranteed connection from train to bus/bus to train at San Luis Obispo.
○ Unreserved service.
▫ Thruway and connecting services.

Services on the Capitol Corridor*
Coaches: Unreserved.
Café: Sandwiches, snacks and beverages.
Bicycles: All trains on the Capitol Corridor are equipped with a limited number of bike racks—passengers may bring bicycles as unboxed carry-on baggage.
- Available first-come, first-served—no reservations or service charge.
- Unboxed bicycles may be put in the bin under connecting Thruway motorcoaches. Amtrak disclaims liability for loss or damage.

Smoking is prohibited entirely on these trains.

The Capitol Corridor is primarily financed through funds made available by California State Department of Transportation. The system is managed by the CCJPA-Capitol Corridor Joint Powers Authority-six local transit agencies in the eight county service area and by BART-The San Francisco Bay Area Rapid Transit District.

No reservations are necessary for travel on the Capitol Corridor. Simply purchase tickets at your nearest staffed Amtrak station or on board if departing from an unstaffed station.

The easy way to and from the San Francisco Bay.
Nobody gives you access to the Bay Area like Amtrak Capitol Corridor®. Thanks to BART regional transit connections at Richmond and Oakland Coliseum, getting to and from the San Francisco and Oakland airports is a breeze. With up to 32 trains a day between Oakland and Sacramento, the best of Northern California is always just a quick train ride away. So hop aboard the Capitol Corridor and discover an easier way around Northern California and the Bay.

Effective date: JANUARY 21, 2008

CAPITOL CORRIDOR®

SAN FRANCISCO — to — RENO

OVER 500 DESTINATIONS

Call 1-800-USA-RAIL
Visit AMTRAK.COM
Experience a journey

SAN FRANCISCO - SAN JOSE
OAKLAND - EMERYVILLE
SACRAMENTO - ROSEVILLE
AUBURN - RENO
And intermediate stations

AMTRAK

The Beaverton Station is the connecting point between the commuter rail and the light-rail systems. This January 2007 view illustrates the street side of the Beaverton facility. *Patrick C. Dorin*

Chapter 8
TRI-MET COMMUTER RAIL: PORTLAND

A feasibility study was started for a commuter rail line in the Washington County area west of Portland in 1996. All of the plans are now in place including new passenger equipment ordered from Colorado RailCar. The new train service is scheduled to begin in September 2008.

The new line is a suburb-to-suburb operation. The route extends northward from Wilsonville to Beaverton, a distance of 14.7 miles over the Portland and Western Railroad. The P&W is a regional freight railroad operation and is a partner in the new commuter rail service. The commuter line will connect with the TriMet light-rail line at Beaverton for transportation to downtown Portland, the airport, the Expo Center, and many other destinations to the east of Portland; as well as to Hillsboro and the region's high-tech corridor to the west.

The Beaverton Transit Center is the north terminal of the commuter line where it will connect with 11 TriMet bus lines and the light-rail MAX Blue and Red lines serving the Beaverton to Hillsboro corridor. The Blue and Red lines provide service west to Hillsboro and east to Portland and the International Airport, plus a connection with the MAX Yellow line to the Expo Center.

Initially there will be five stations on the commuter line with a total of about 800 Park and Ride spaces at four of the stations. The four stations are as follows:

The Washington Square Station is within walking distance of the Washington Square Mall and will have about 160 parking spaces. It is also a connection to local TriMet bus lines.

The Tigard Transit Center Station is located in downtown Tigard and will have 120 parking spaces. The station also has connections with five TriMet bus lines.

The Tualatin Station will have 120 parking spaces and connections with a local TriMet bus service.

The Wilsonville Station is at the south end of the 14.7-mile route and will have 400 parking spaces. The station is also a connection with SMART buses serving the residential and employment areas.

Substantial new growth and redevelopment is expected at most of the five station areas similar to the new developments that have occurred at the MAX station areas.

The stations will also be decorated with art and sculpture using imagery distinct for each locale. The Commuter Rail Art Advisory Committee selected artists Frank Boyden and Brad Rude to develop the artwork for the stations.

The use of the existing rail route of the Portland and Western substantially reduced development and investment costs. A short extension from the P&W trackage constructed on Lombard Avenue between Farmington Road and the Beaverton Transit Center will provide easy and direct "cross the platform" connections between the commuter trains and the light-rail and bus systems.

The project ridership between 2008 and 2020 is expected to grow to 3,000 to 4,000 passengers per week-day. However, it is interesting to note that the projected ridership for rail systems is usually exceeded.

The initial train service levels will be during the rush hour periods with a train every 30 minutes. The running time between Beaverton and Wilsonville is about 27 minutes. This is about 50% of the time required for driving. Furthermore, passengers en route to downtown Portland will have only a 15- to 20-minute ride to their destination stop. Thus passengers will be able to travel to downtown Portland in about 50 minutes compared to nearly 90 minutes of driving depending upon traffic congestion and weather conditions.

The passenger equipment will be the new design of Diesel Multiple Units built by Colorado Railcar, known as DMUs. With the exception of changing the number of cars for a particular train, the DMUs require very little switching and are bi-directional. The equipment does not need to be turned around at the final terminals of Wilsonville and Beaverton.

The new commuter train service will be adding a new level of public transportation for the Portland area, particularly since the planning includes coordination with the light-rail and bus systems. It is interesting to note that bus patronage increases when it is operated in conjunction with rail transit connections. The Portland area is to be commended for their work to create a balanced public transportation system. There is no doubt that there is more to come in future planning for both commuter services and the light-rail system in the City of Roses.

The Washington County Commuter Rail map illustrates the route for the new service between Beaverton and Wilsonville. *Tri-Met and Washington County*

The website for the Tri-Met: *www.trimet.org*

This is the type of equipment on order as of 2007 for the new commuter rail service. The Diesel Multiple Unit type of equipment can be very effective for the new type of service for the Portland area. *Portland Tri-Met*

This view shows the Tri-Met trackage for the light-rail system, and the area where passengers will be able to easily transfer between the two rail systems at Beaverton. *Patrick C. Dorin*

These two views show the Hatfield Government Center station at the west end of the Tri-Met System at Hillsboro. The stations are friendly and, in this case, the trains are ready to depart for Portland and Greshem, which is located on the far eastern end of the main line route of Tri-Met. Thus when passengers transfer at Beaverton to the Light Rail, they have choices for the Airport, downtown Portland, west to Hillsboro and as far east as Greshem. Passengers are also able to transfer at other station locations for trains to the Expo Center. Tri-Met is a very convenient way for Portland area travel, and the system has several plans for expanding the rail routes and new bus connections. It is a balanced approach and many other metro areas should take a closer look at the Portland systems. *Patrick C. Dorin*

The Portland Station is an important stop in downtown for the light-rail system and buses, and for providing a connection to the Beaverton Station. *Patrick C. Dorin*

Washington County Commuter Rail

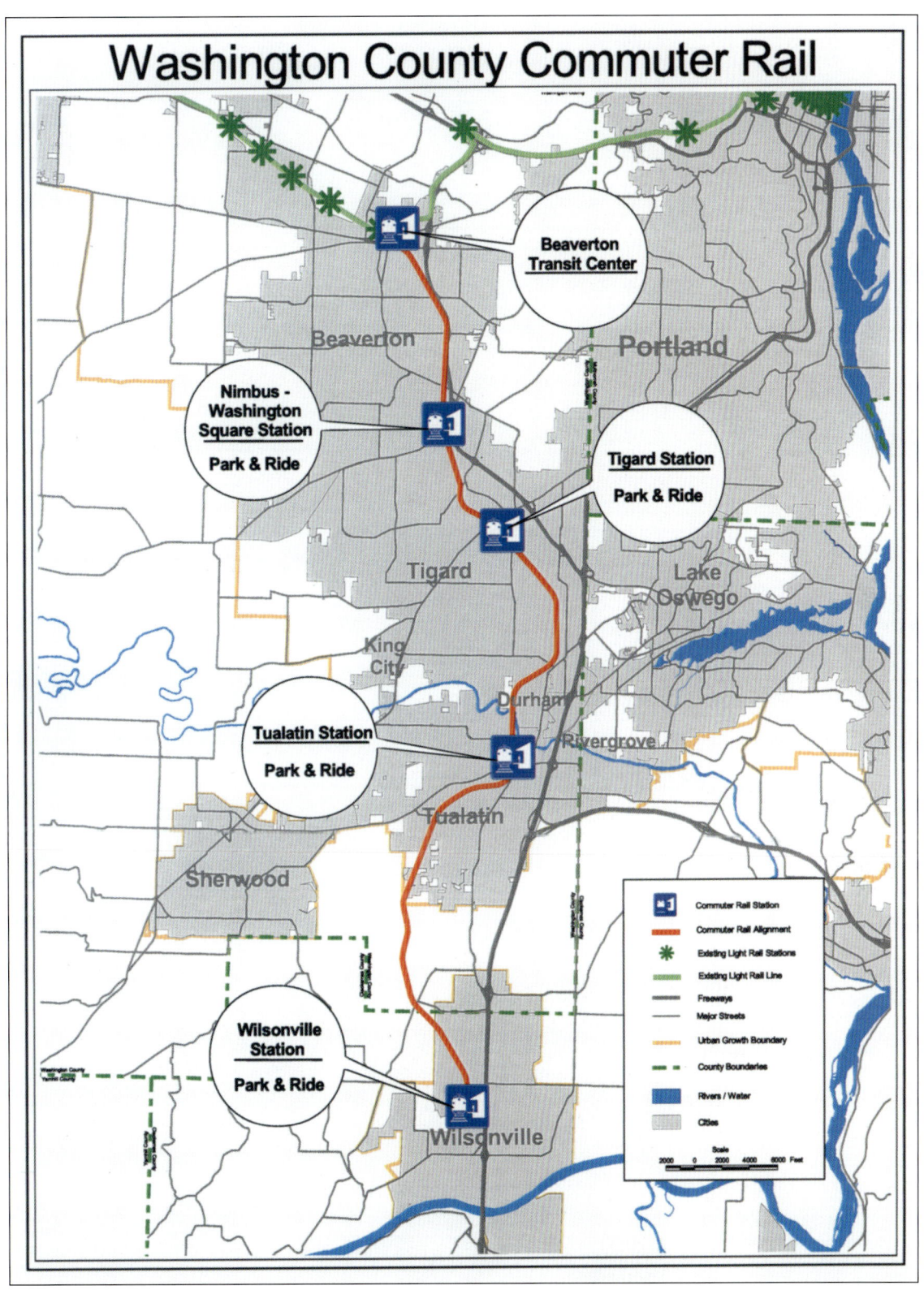

A Sounder train is pausing at the Tacoma Dome Station. This is a multi-modal terminal for both rail and bus services plus extensive parking facilities. Sounder trains operate in the push mode to Seattle, and the pull mode to Tacoma. The station platforms at Tacoma are designed to smoothly handle the movement of passengers for boarding and departing from the trains. *Sound Transit*

Chapter 9
THE SOUNDER: SEATTLE

The Sounder commuter train services operate between Tacoma and Seattle, and north of Seattle to Everett over the BNSF. This is the former Great Northern route north of Seattle, and the route that was originally the Northern Pacific (with Great Northern-granted trackage rights) to Tacoma. The distance from Seattle to Tacoma is 40 miles, and the Seattle–Everett route is 33 miles. Plans include the extension of the Sounder commuter from Tacoma to Lakewood, a distance of 9 miles. The new service is currently scheduled to begin operation in 2011.

The Sounder service began between Tacoma and Seattle in September 2000. The initial service was two rush hour trains from Tacoma to Seattle in the morning, and two trains southbound in the evening. The running time between Seattle and Tacoma was 59 or 60 minutes depending upon the train schedule.

The Sounder train service was extended from Seattle to Everett in December 2003 with one rush hour train from Everett in the morning, and one from Seattle in the afternoon. By 2005, Sounder added a second rush hour train to Seattle in the morning, and to Everett in the afternoon. A third rush hour trip to Seattle in the morning, and to Everett in the afternoon was added in 2007. As of this writing in October 2007, the Everett service has three rush hour round-trips Monday through Friday. An additional train is planned to go into service during the year 2008.

Another service innovation on the Everett route was implemented in the year 2004. Rail Plus provides an opportunity for Sounder passengers, who have the appropriate monthly or annual passes, to use Amtrak trains stopping at Everett, Edmonds and Seattle. There are presently (fall 2007) two Amtrak trains (the Cascades) operating in each direction that can take Sounder passengers at the three stations listed previously. Thus passengers have a choice of five trains in each direction Monday through Friday. Amtrak tickets and Sounder passes are honored on both rail services. Here is another example of the potential co-operative service that can be provided with Amtrak and a commuter rail system.

Train service on the Seattle–Tacoma route has increased since 2000 when operations began with the two rush hour trains. By 2006, there were four trains to Seattle in the morning and four to Tacoma in the evening, Sound Transit added more service on the route in 2007 including a reverse direction rush hour train to Tacoma in the morning, and to Seattle in the afternoon. As of the fall 2007, passengers have a choice of five rush hour trains to Seattle in the morning, and five to Tacoma in the afternoon; plus the rush hour train to Tacoma in the morning and to Seattle in the afternoon.

This adds up to a total of six Sounder trains in each direction for a total of twelve trains. Adding the three Sounder trains in each direction on the Everett route brings the total number of trains to eighteen per weekday.

The reverse direction commuter run on the Seattle–Tacoma route is becoming more common in North America. It is one more example of the interdependency between the cities and the surrounding suburbs.

Additional train service is also provided for special events such as selected Mariner and Seahawk home games.

It is anticipated that there will eventually be 18 trains serving the Lakewood–Tacoma–Seattle route. This will include rush hour trains in each direction during the morning and evening periods.

Ridership since 2000 has continued to grow moving through the first decades of the 21st century. The annual ridership figures are as follows:

Annual Ridership Figures

Year	Total Passenger Count	Average Daily Counts
2000	102,552	1,227
2001	562,740	2,135
2002	672,495	2,514
2003	751,163	2,794
2004	955,298	3,459
2005	1,267,973	4,564
2006	1,692,971	6,310

Motive Power and Passenger Equipment

The Sounder Commuter trains are equipped with Bi-level Push-Pull cars built by Bombardier in Thunder Bay, Ontario. Some of the equipment had been loaned out to other commuter carriers prior to going into service on the Sounder Routes. Just one example was the test run of the new commuter train service on the North Star Line between Minneapolis and Big Lake, Minnesota.

The motive power roster as of 2006

Type	Number Series	Built	Builder
F59	900—911	2000	GM

As of this writing in the fall of 2007, No. 901 is operating in Southern California on the Metrolink, and the 902 has been loaned to the Virginia Railway Express.

The Bi-level passenger equipment is painted with a neat color scheme depicting the waves on Puget Sound. The fleet is well-liked by the passengers, and the trains themselves have been a drawing card for new riders. This is in part true because of the comfort level of the Bombardier equipment, not to mention the increasing traffic congestion in the Pacific Northwest. All of the equipment was built by Bombardier.

Sounder passengers have a scenic route for traveling to and from work, shopping and other activities. This train is passing Carkeek Park along the Puget Sound. *Sound Transit*

The passenger equipment roster is as follows:

Passenger Equipment

Type	Number Series	Built
Cab Coaches	101 to 111	2000—2003
	301 to 307	2000—2003
Coaches	201 to 218	2000—2002
	227 and 228	2000—2002
	231 to 240	2000—2002
	401 to 410	2000—2002

As of this writing in 2007, the following passenger cars have been leased to Metrolink and Virginia Railway Express:

VRE: 102, 103, 201, 204, 206, and 212

Metrolink: 104, 106, 107, 108, 210, 213, 231, 237, 238, 239, and 240

The leasing of equipment can provide model railroaders with ideas for commuter train modeling. For example, one could have a new commuter route starting up and have leased both motive power and passenger equipment from other carriers for the operations. One could even have a mix of equipment from different commuter railroads.

The Sounder Website: *www.soundtransit.org*

This southbound train is rolling at track speed just south of Carkeek Park after having departed Seattle with a four-car consist. The beautiful Puget Sound is to the left. *Sound Transit*

This is the Everett Station, which is not only a multi-modal transit terminal, but includes the University of Washington as part of the station complex. The station area has parking space for not only passengers, but also for the university students and staff. Furthermore, with the new commuter train services, and the pooling ticket arrangements with Amtrak's Cascades, people have greater flexibility traveling to and from school. *Sound Transit*

A four-car Sounder commuter train, powered by No. 910, rolls along on the double-track main line between Seattle and Tacoma. *Jeff Koeller*

The same train is now rolling by the Mukilteo station sign in a very scenic area with pleasant hills and beautiful residential areas. *Jeff Koeller*

This Sounder train is pausing on its southbound run to load passengers for the rush hour trip to Seattle. The Seattle–Everett trains had but one stop en route at Edmonds. *Jeff Koeller*

A four-car train is moving into the Seattle Station and will soon be ready for its southbound rush hour trip to Tacoma as train No. 1501, here in June 2005. *Patrick C. Dorin*

A few minutes later, the equipment for train No. 1503 has arrived at the Seattle station. This photo shows the two sets of equipment prior to departure for Tacoma. *Patrick C. Dorin*

This view shows the Seattle station platforms and part of the facilities in June 2005. *Patrick C. Dorin*

Sounder Coach No. 111. *Patrick C. Dorin*

Sounder Coach No. 408. *Jeff Koeller*

The Cab Cars have a window on each side as illustrated here with Sounder No. 305. *Jeff Koeller*

The motive power, such as No. 910, illustrates the waves of Puget Sound from the front of the engine and along the sides. *Jeff Koeller*

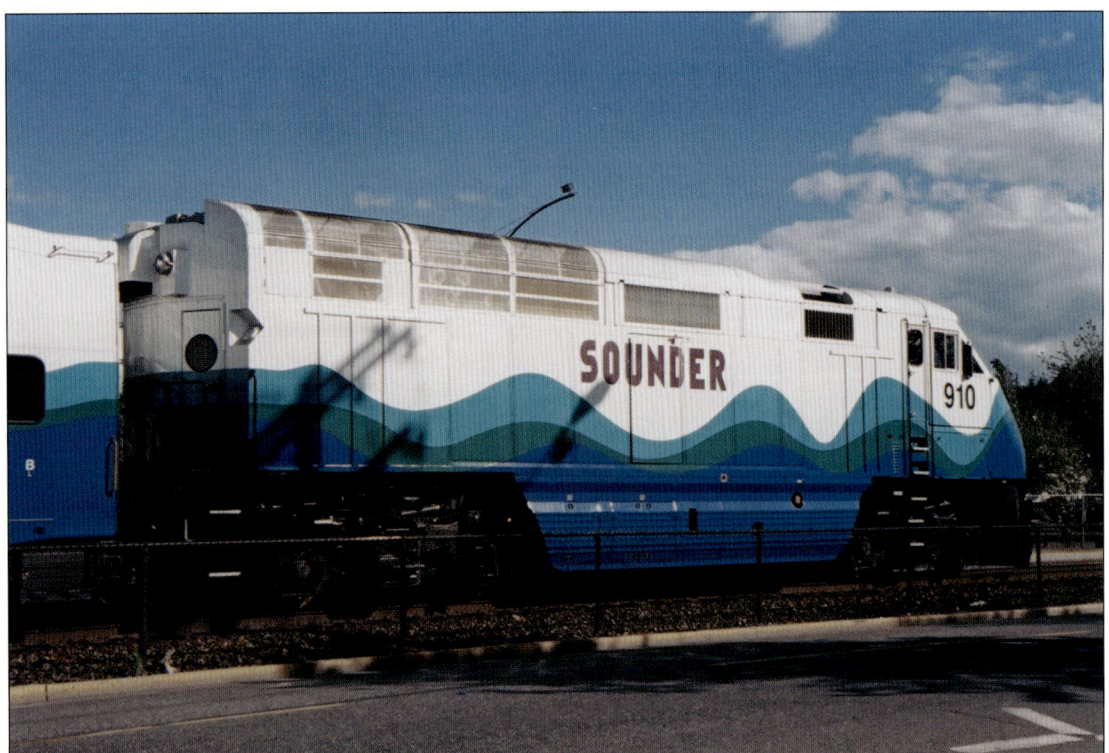

The waves continue and wrap around the rear of the Sounder motive power. *Jeff Koeller*

The décor for No. 911 is for the Seattle Seahawks. The 911 is powering a seven-car train which is about to depart for Tacoma in this June 2005 photo. *Patrick C. Dorin*

The décor includes an ad as to where to call for tickets for a game, plus information on how to log on to the Sounder web site for information on train schedules for the games. *Patrick C. Dorin*

The joint ticketing arrangement for the commuter service includes service on Amtrak's Cascades in the Sounder Service Area. The Cascades have been equipped with Talgo train sets. Refer to the timetables in this chapter for the Cascade schedules and the pool service, known as the Rail Plus Partnership, with Sounder. The set of Cascade equipment is at the Portland Station in April 2007. *Patrick C. Dorin*

Tacoma - Seattle · NORTHBOUND

TRAIN #	TACOMA	PUYALLUP	SUMNER	AUBURN	KENT	TUKWILA	SEATTLE
1500	5:00 a.m.	5:12 a.m.	5:17 a.m.	5:26 a.m.	5:34 a.m.	5:41 a.m.	6:00 a.m.
1502	5:45 a.m.	5:57 a.m.	6:02 a.m.	6:11 a.m.	6:19 a.m.	6:26 a.m.	6:45 a.m.
1504	6:20 a.m.	6:32 a.m.	6:37 a.m.	6:46 a.m.	6:54 a.m.	7:01 a.m.	7:20 a.m.
1506	6:50 a.m.	7:02 a.m.	7:07 a.m.	7:16 a.m.	7:24 a.m.	7:31 a.m.	7:50 a.m.
1508	7:20 a.m.	7:32 a.m.	7:37 a.m.	7:46 a.m.	7:54 a.m.	8:01 a.m.	8:20 a.m.
1510	4:45 p.m.	4:57 p.m.	5:02 p.m.	5:11 p.m.	5:19 p.m.	5:26 p.m.	5:45 p.m.

Seattle - Tacoma · SOUTHBOUND

TRAIN #	SEATTLE	TUKWILA	KENT	AUBURN	SUMNER	PUYALLUP	TACOMA
1501	6:10 a.m.	6:26 a.m.	6:33 a.m.	6:40 a.m.	6:50 a.m.	6:54 a.m.	7:10 a.m.
1503	3:35 p.m.	3:51 p.m.	3:58 p.m.	4:05 p.m.	4:15 p.m.	4:19 p.m.	4:35 p.m.
1505	4:20 p.m.	4:36 p.m.	4:43 p.m.	4:50 p.m.	5:00 p.m.	5:04 p.m.	5:20 p.m.
1507	4:45 p.m.	5:01 p.m.	5:08 p.m.	5:15 p.m.	5:25 p.m.	5:29 p.m.	5:45 p.m.
1509	5:15 p.m.	5:31 p.m.	5:38 p.m.	5:45 p.m.	5:55 p.m.	5:59 p.m.	6:15 p.m.
1511	5:55 p.m.	6:11 p.m.	6:18 p.m.	6:25 p.m.	6:35 p.m.	6:39 p.m.	6:55 p.m.

Seattle - Tacoma Single trip ticket fares

		TUKWILA	KENT	AUBURN	SUMNER	PUYALLUP	TACOMA
Seattle	Adult	$3.25	$3.50	$3.75	$4.25	$4.25	$4.75
	Youth	$2.25	$2.50	$2.75	$3.00	$3.00	$3.50
	Sr/Disabled†	$1.50	$1.75	$1.75	$2.00	$2.00	$2.25
Tukwila	Adult		$2.75	$3.25	$3.50	$3.75	$4.00
	Youth		$2.00	$2.25	$2.50	$2.75	$3.00
	Sr/Disabled†		$1.25	$1.50	$1.75	$1.75	$2.00
Kent	Adult	$2.75		$2.75	$3.25	$3.50	$3.75
	Youth	$2.00		$2.00	$2.25	$2.50	$2.75
	Sr/Disabled†	$1.25		$1.25	$1.50	$1.75	$1.75
Auburn	Adult	$3.25	$2.75		$3.00	$3.00	$3.50
	Youth	$2.25	$2.00		$2.25	$2.25	$2.50
	Sr/Disabled†	$1.50	$1.25		$1.50	$1.50	$1.75
Sumner	Adult	$3.50	$3.25	$3.00		$2.75	$3.00
	Youth	$2.50	$2.25	$2.25		$2.00	$2.25
	Sr/Disabled†	$1.75	$1.50	$1.50		$1.25	$1.50
Puyallup	Adult	$3.75	$3.50	$3.00	$2.75		$3.00
	Youth	$2.75	$2.50	$2.25	$2.00		$2.25
	Sr/Disabled†	$1.75	$1.75	$1.50	$1.25		$1.50
Tacoma	Adult	$4.00	$3.75	$3.50	$3.00	$3.00	
	Youth	$3.00	$2.75	$2.50	$2.25	$2.25	
	Sr/Disabled†	$2.00	$1.75	$1.75	$1.50	$1.50	

† When paying a senior or disabled fare, a regional reduced fare permit is required. Medicare card holders are eligible for a permit.

Everett - Seattle · SOUTHBOUND

TRAIN #	EVERETT	EDMONDS	SEATTLE
1701	6:12 a.m.	6:36 a.m.	7:10 a.m.
1703	6:42 a.m.	7:06 a.m.	7:40 a.m.
1705	7:12 a.m.	7:36 a.m.	8:10 a.m.
Amtrak 513*	9:55 a.m.	10:21 a.m.	10:55 a.m.
Amtrak 517*	8:54 p.m.	9:19 p.m.	10:05 p.m.

Seattle - Everett · NORTHBOUND

TRAIN #	SEATTLE	EDMONDS	EVERETT
Amtrak 510*	7:40 a.m.	8:07 a.m.	8:31 a.m.
1700	4:33 p.m.	5:00 p.m.	5:31 p.m.
1702	5:05 p.m.	5:32 p.m.	6:03 p.m.
1704	5:35 p.m.	6:02 p.m.	6:33 p.m.
Amtrak 516*	6:40 p.m.	7:07 p.m.	7:31 p.m.

*Rail Plus Trains 510, 513, 516 and 517 are operated by Amtrak and are part of the Rail Plus Program serving Seattle, Edmonds and Everett stations, but **only accept monthly passes valid on Sound Transit services.** Pass upgrades, Day Passes and Single-Trip tickets are not accepted by Amtrak. Amtrak trains board on the west side of Weller Bridge. See www.soundtransit.org for more information.

Everett - Seattle Single trip ticket fares

		MUKILTEO**	EDMONDS	SEATTLE
Everett	Adult	$2.75	$3.50	$4.50
	Youth	$2.00	$2.50	$3.25
	Sr/Disabled†	$1.25	$1.75	$2.25
Mukilteo** (2008)	Adult		$3.25	$4.00
	Youth		$2.25	$3.00
	Sr/Disabled†		$1.50	$2.00
Edmonds	Adult		$3.25	$3.50
	Youth		$2.25	$2.50
	Sr/Disabled†		$1.50	$1.75

** Service to Mukilteo Station is scheduled to begin in Spring 2008.

EVENT SERVICE: The Sounder will run to select Seahawks games. Visit us online for more information.

HOLIDAY SERVICE: No Sounder service on Veterans Day, Thanksgiving Day, Christmas Day and New Years Day.

RIDER INFORMATION: 1-888-889-6368 / 1-888-713-6030 TTY / www.soundtransit.org

Children: Children under six ride free with a fare paying passenger.
Youth: age 6 to 18
Adult: age 19 to 64
Senior/Disabled†: age 65 or older, qualifying disability

This page from the Sounder Commuter Rail Schedule (September 24, 2007, through February 8, 2008) illustrates the train services both north and south of Seattle. Trains 1501 and 1510 illustrate the reverse commuter rush hour schedules between Tacoma and Seattle. The RAIL PLUS TRAINS 510, 513, 515, and 517 are operated by Amtrak. The notes below the Seattle-Everett schedules show that only Sounder passengers with monthly passes are valid on the Amtrak trains. *Sound Transit*

 ## Sounder commuter rail

Sound Transit's commuter rail operates Monday through Friday in the peak direction, at peak hours. There are four morning trips leaving Tacoma traveling northbound to Seattle and two morning trips leaving Everett and traveling southbound to Seattle. In the afternoon, there are four trips departing Seattle heading to Tacoma and two trips departing Seattle heading to Everett. Amenities include worktables and computer plug-ins, fully accessible restrooms in each car and cup holders. Please see the schedule at right for the most up-to-date Sounder schedule.

 **RailPlus promotion**
Through the Rail Plus partnership, Sound Transit and Amtrak Cascades have expanded commuter rail service in north Puget Sound by offering more options for your weekday peak trips between Everett and Seattle. With this partnership, Amtrak Cascades serves Sounder stations at King Street, Edmonds and Everett. Paying your fare is easier too. Monthly PugetPass, U-Pass and FlexPass cards are now accepted onboard the Amtrak Rail Plus trains and valid Amtrak tickets may be used on Sounder to travel between Everett and Seattle. See below for more Sounder and Amtrak fare information.

The pass you use on Amtrak must be equal to the full-fare price of the Sounder trip. Although accepted on Sounder trains, pass upgrades, Day Passes or Single-Trip tickets are not accepted on Amtrak.

Amtrak tickets are only accepted on weekday Sounder service between Seattle and Everett and are not valid on event service trains. Reduced Sounder fares are available for youth, senior and disabled riders. Children under six do not ride free on Amtrak and riders with bicycles must pay additional fees.

Everett to Seattle
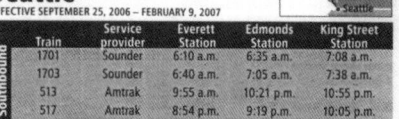

Tacoma to Seattle
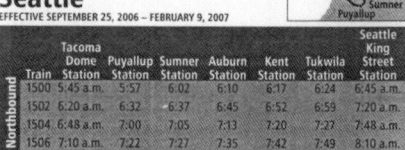

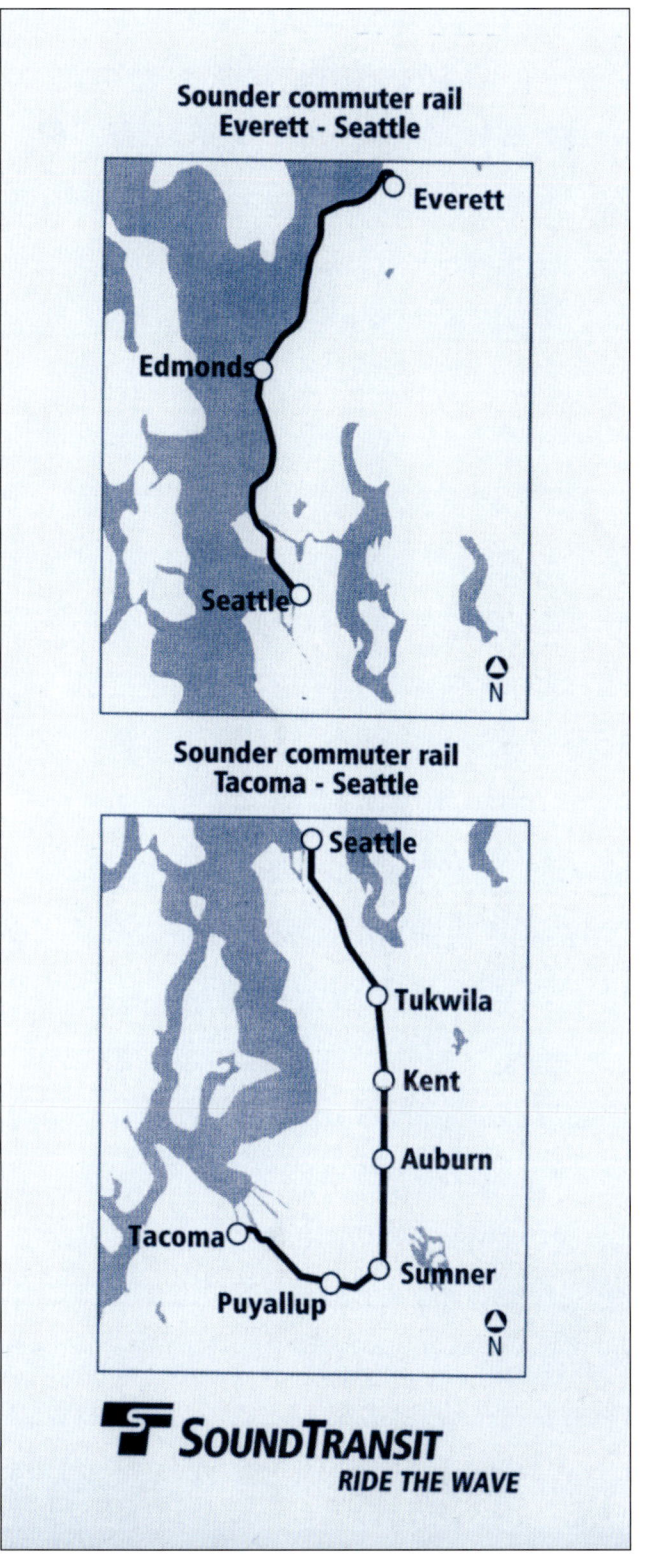

These two maps illustrate the routes between Seattle and Everett, and between Seattle and Tacoma. *Sound Transit*

The rush hour trains have consists of up to seven cars, such as in this scene of an eastbound train rolling through the beautiful scenery of British Columbia just west of Port Henry near Maple Ridge. *WCE Photo*

Chapter 10
THE WEST COAST EXPRESS: VANCOUVER

The West Coast Express is among the newest commuter rail services in Canada and the United States. The WCE provides train service from Mission City to downtown Vancouver, a distance of 42.1 miles over the Canadian Pacific Railway. The train service began on November 1, 1995. The West Coast Express carried over 18 million riders during its first decade of operation through the end of 2005. Weekday ridership has grown from 5,000 per day in the 1990s, to over 9,000 per day as we move through the year 2006.

The trains are operated Monday through Friday only with five rush hour schedules to Vancouver in the morning, and five trains to Mission City in late afternoon and early evening. The consist of each train ranges from four to nine cars powered by one locomotive. There is additional bus service over the route.

One of the interesting aspects of the West Coast Express is their HOMERUN Emergency Daytime Travel Insurance. Passengers who suddenly have a situation where they have to get home, and have the insurance plan, can easily call the WCE and secure a ride home by taxi. The HOMERUN travel insurance pays the fare up to three times per year. Without mid-day train schedules, passengers with the insurance have a new level of security in the event of a medical or family emergency.

Motive Power and Passenger Equipment

The West Coast Express has five F59PHI, numbered 901 to 905. They were built in 1995, The 906, type MP36PH-3C arrived in 2006.

The Bi-level push-pull passenger equipment was built by Bombardier in Thunder Bay, Ontario. Passenger seating capacity is as follows:

Cab Coach Cars handling Wheel Chairs and Bicycles	136
Cab Coach Cars without Wheel Chair and Bicycle Racks	142
Trailer Coaches handling Wheel Chairs and Bicycles	142
Trailer Coaches without Wheel Chair and Bicycle Racks	148

The air-conditioned coaches are equipped with computer plug-ins, work tables, armrests, drink holders and washrooms. The trains also have on-board coffee service, which is very convenient for the passengers. The WCE originally ordered 28 cars, of which 8 were cab cars and 20 were trailers. The total number of cars as we move into 2007 is 36.

Type	Number Series
Cab-Coach	101—108
Trailer Coaches	201—220 and 301—308
	Cars 201 to 205 include a Cappuccino Bar.

The website for the West Coast Express: *www.westcoastexpress.com*

The following photos, timetables and other illustrations provide an overall coverage of this new commuter train service, which can serve as an example for many other cities and suburban areas throughout North America.

The sun is reflecting off of the motive power as an eastbound evening train heads for the eastern terminal at Mission City. The train is just west of Coquitlam Central. *WCE Photo*

Motive power and trains are lined up at the Vancouver Waterfront Station with the city's picturesque skyline in the background. *WCE Photo*

A morning rush hour train is pausing at the Port Haney station for passengers while en route from Mission City to Vancouver. Passengers have said it is the most relaxing way to travel to work, and provides new levels of energy for the day ahead. Passengers can sip cappuccino, and sit back and relax and enjoy the scenery while traveling either to work or home. *WCE Photo*

more information

customer service

By phone: (604) 488-8906 • Fax: (604) 682-0562
24-hour automated: (604) 683-RAIL
1-800-570-7245
Online: **www.westcoastexpress.com**
Email: wcecustomer_service@translink.bc.ca
In person: Waterfront Station
Suite 295-601 West Cordova St.
Vancouver BC V6B 1G1

connection info

Our Customer Service Representatives will
be pleased to provide connection information to other TransLink services. Or you
may contact the following services directly.
TransLink: (604) 953-3333
(includes transit, SkyTrain & SeaBus)
handyDART: (604) 953-3400 information only. (8:00am-4:00pm)
Central Fraser Valley Transit System: (604) 854-3232
Morning handyDART service in Mission: (604) 820-2433

gentle reminders

No Smoking: Smoking is not permitted on trains,
the platform, or in the station house.
Proof of Payment: Be prepared to show a valid
WCE ticket/pass at all times – failure to do so can
result in fines and suspension of ridership privileges.
Be Considerate: Please monitor the volume of your
conversation and leave seats and aisles clear of
personal belongings.

WEST COAST EXPRESS
we go the distance.
West Coast Express is an operating subsidiary of TransLink.

passenger's guide

Your Guide to Comfort
and Convenience.

The July 2001 edition of the "Passenger's Guide" is one example of brochures that outlines the services provided by the West Coast Express including the schedules for the Morning and Evening rush hour services.

service & schedule

providing a link

West Coast Express (WCE) is a premium,
long haul commuter service operating
between Mission and Vancouver. As
an operating subsidiary of TransLink,
all WCE tickets are transferable onto
SeaBus, SkyTrain, and transit buses.

take the rush out of rush hour

Leave your car at home and enjoy the
ride aboard WCE. The seats are comfy
and the coffee's great. Whether you're
catching-up on work or just relaxing
with the paper, everything's on board
to ensure you have a comfortable and
convenient commute.

comfort and convenience – up and down the line

To make your ride more enjoyable,
we offer an assortment of beverages
and snacks from our cappuccino car.
We also provide work stations complete
with computer plug-ins. Plus, each
train has a pay-per-use mobile phone.

peak-hour performance

WCE provides regular service during peak
hours Monday through Friday (except
statutory holidays). Weekday service consists
of five morning (westbound) and five
afternoon (eastbound) trains.

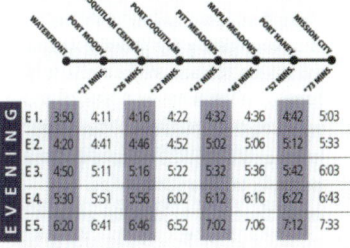

MORNING	MISSION CITY	PORT HANEY *73 MIN.	MAPLE MEADOWS *56 MIN.	PITT MEADOWS *50 MIN.	PORT COQUITLAM *44 MIN.	COQUITLAM CENTRAL *36 MIN.	PORT MOODY *30 MIN.	WATERFRONT *25 MIN.
W1.	5:27	5:44	5:50	5:54	6:04	6:10	6:15	6:40
W2.	5:57	6:14	6:20	6:24	6:34	6:40	6:45	7:10
W3.	6:27	6:44	6:50	6:54	7:04	7:10	7:15	7:40
W4.	6:57	7:14	7:20	7:24	7:34	7:40	7:45	8:10
W5.	7:27	7:44	7:50	7:54	8:04	8:10	8:15	8:40

*Estimated time to Vancouver

EVENING	WATERFRONT	PORT MOODY *21 MIN.	COQUITLAM CENTRAL *26 MIN.	PORT COQUITLAM *33 MIN.	PITT MEADOWS *47 MIN.	MAPLE MEADOWS *56 MIN.	PORT HANEY *67 MIN.	MISSION CITY *73 MIN.
E1.	3:50	4:11	4:16	4:22	4:32	4:36	4:42	5:03
E2.	4:20	4:41	4:46	4:52	5:02	5:06	5:12	5:33
E3.	4:50	5:11	5:16	5:22	5:32	5:36	5:42	6:03
E4.	5:30	5:51	5:56	6:02	6:12	6:16	6:22	6:43
E5.	6:20	6:41	6:46	6:52	7:02	7:06	7:12	7:33

*Estimated time from Vancouver

The pre-commuter rail service on the Alaska Railroad will operate with existing passenger equipment. A typical example of the types of coaches operated by the Alaska Railroad is a streamlined car for intercity service. Coach No. 206 was built by a Korean car builder and includes larger windows for better visibility of the Alaskan scenery. The 206 is part of the consist of a train in July 2000 about to depart Anchorage for Seward. *Patrick C. Dorin*

Chapter 11
PROPOSED COMMUTER RAIL ANCHORAGE, ALASKA

The Alaska Railroad is a full service railroad in the United States with both freight and passenger service. Passenger trains operate north and south of Anchorage connecting Seward, Whittier, Talkeetna, Denali National Park and Fairbanks. A proposed regional commuter rail service between the Matanuska-Susitna Valley, Anchorage, and Girdwood has community interest and continues to develop.

Anchorage serves the regional employment center for a fifty-mile radius and is the home of a number of major employers including oil companies, finan-

cial institutions, retail centers, and federal and state government offices. Anchorage is served by a single highway subject to morning and evening rush hour congestion. Severe weather in winter, frequent accidents, and wildlife encounters are great causes of concern.

The Alaska Railroad has been cooperating with the local interests in developing the infrastructure backbone of a future commuter rail service. Intermodal facilities in Palmer and Wasilla have either been constructed or are in the early planning stages. Facil-

Alaska Railroad GP40-2, No. 3007, here in July 2000, is typical of the power operated for the passenger services. A power car is part of the train consist since the GP40-2 does not contain a head-end power component. *Patrick C. Dorin*

The following map illustrates the route of the proposed new service. Photos illustrate the color scheme of the Alaska Railroad and some of the equipment presently in operation for the passenger services.

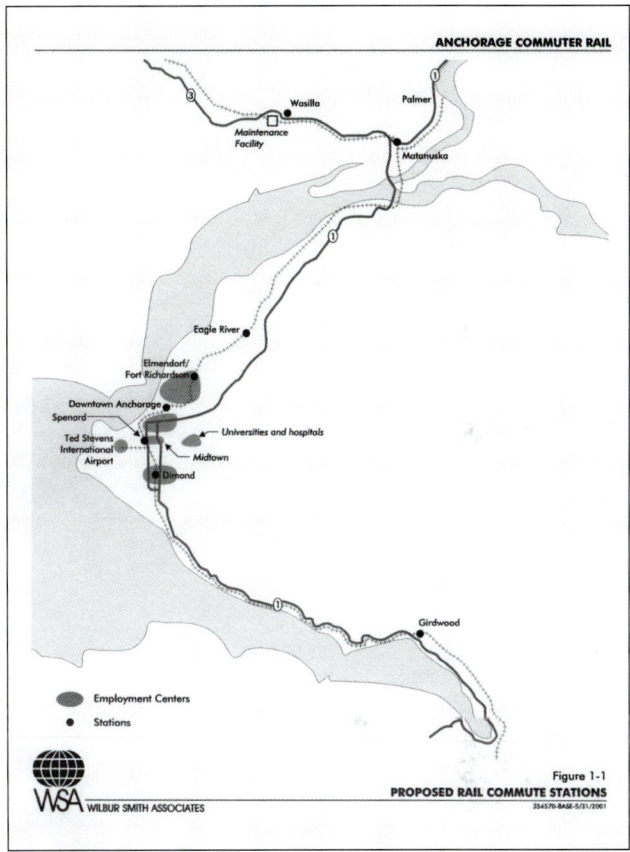

Alaska Railroad Map
The proposed commuter rail service would operate between Anchorage and Wasilla and Palmer to the north, and between Anchorage and Girdwood to the south. The milepost numbers reflect the mileage from Seward at milepost 0.0. Girdwood is at 74.5, while Anchorage is 114.3 and Wasilla at 159.8. The milepost number for Palmer is 6.5, which is the length of the Palmer Branch. *Alaska Railroad*

ities in Anchorage are undergoing major renovations to meet future long-distance passenger train service as well as a local commuter train service. The Ted Stevens Anchorage International Airport is home to the Bill Sheffield Alaska Railroad Station already serving as a regional rail transportation center. Intermodal facilities at the Dimond Shopping Center, a regional retail and commercial center, are in the early planning stages.

The full "build-out" stage will include service south to Girdwood, a distance of 30 miles, and to the Matanuska-Susitna Valley including Wasilla and Palmer, both north of Anchorage approximately 40 miles away. The "Valley" serves as a bedroom community to Anchorage and local military bases. This area serves as the primary generator of commuter traffic into the Anchorage bowl area.

While the physical facilities of a potential commuter rail service have been adequately identified, the actual implementation is dependent on a number of factors out of the Alaska Railroad's control. The region includes three different local governments, one separate ferry system, and two separate bus transit services. A regional transit authority will have to be implemented which includes all of the above as well as the commuter rail service. At this time, 2007, there is no local source of revenue to support such enhanced transit service.

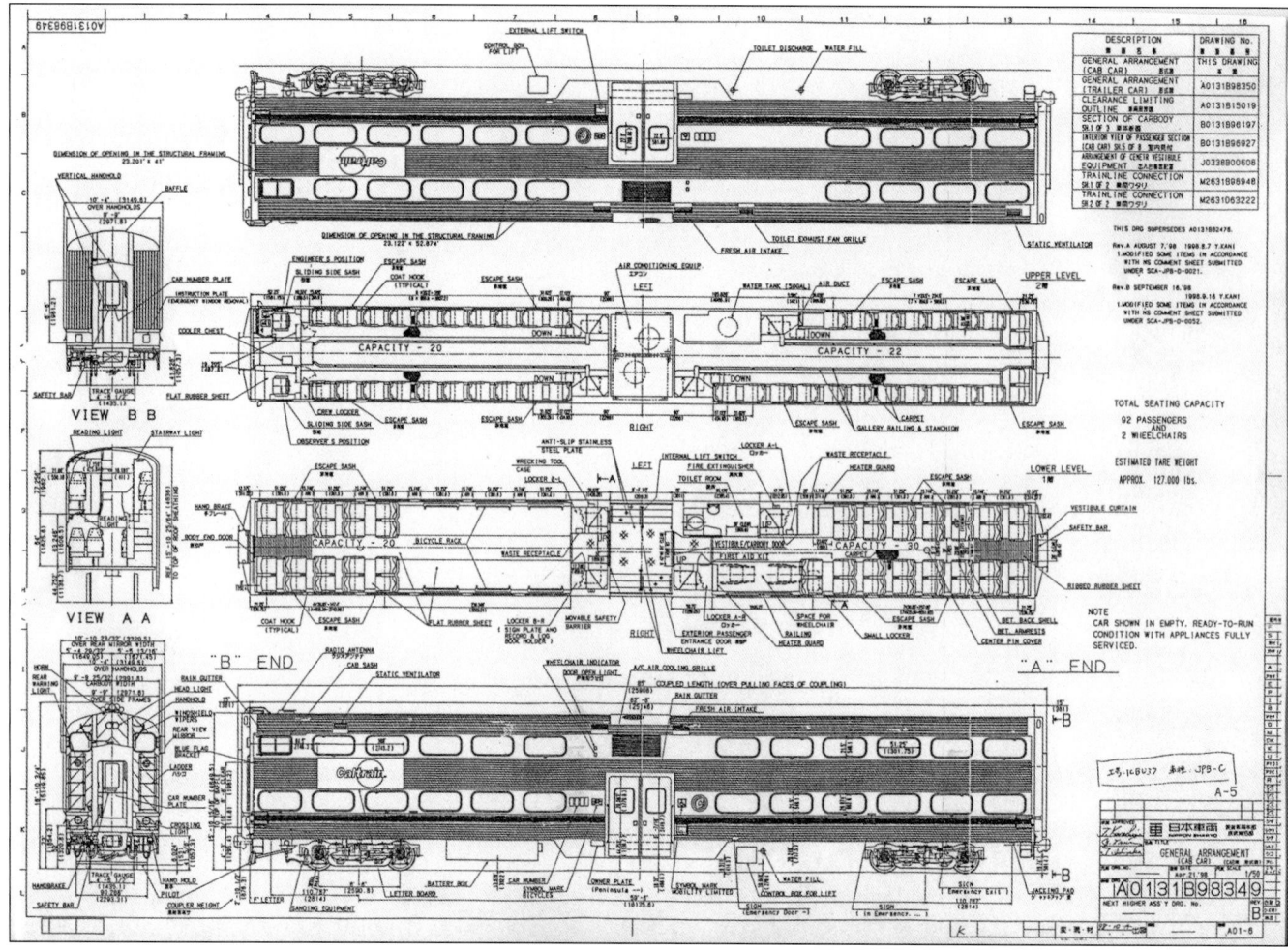

Cab Car Bi-level Diagram. *Nippon Sharyo*

Epilogue
COMMUTER RAIL PASSENGER EQUIPMENT

Bombardier has produced most of the passenger equipment on the new commuter rail lines along the Pacific Coast. All of the commuter rail systems, with the exception of Portland, have the Bombardier Bi-level commuter coaches in operation. Nippon Sharyo was the builder of gallery equipment for the Caltrain services that replaced the original Southern Pacific gallery cars. Caltrain is the only operation with two different types of commuter coaches. The new service for Portland, Oregon, has Diesel Multiple Unit

equipment on order from Colorado Railcar. All of the equipment operates in the Push-Pull concept for the train services.

Following are drawings or diagrams of the commuter passenger equipment in service on the West Coast of North America.

The Nippon Sharyo USA, Inc. diagrams illustrate the coach and cab coach design in operation for Caltrain between San Francisco, San Jose and Gilroy.

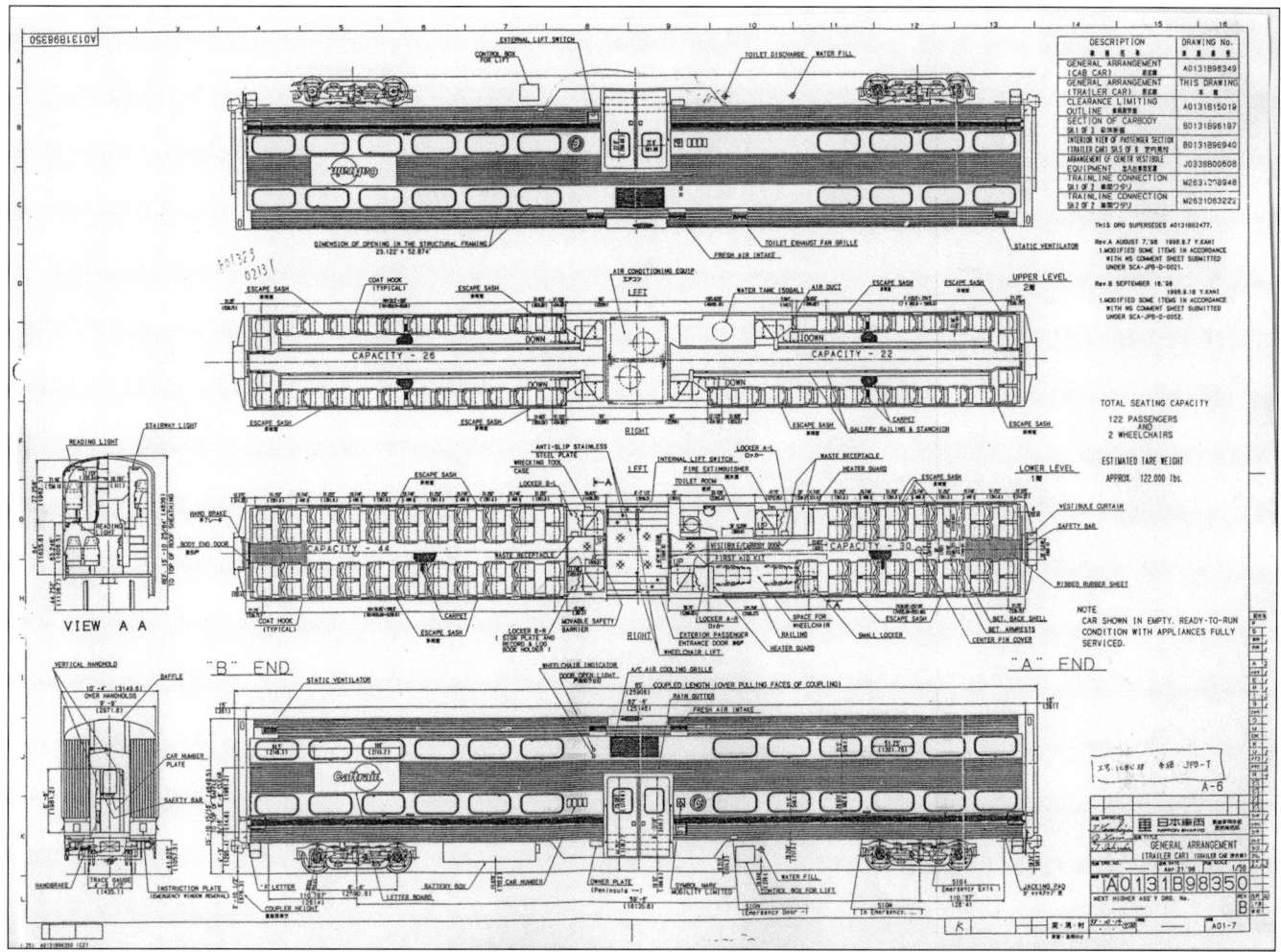

Bi-level Coach Diagram. *Nippon Sharyo*

Colorado Railcar is in the process of building the single level Diesel Multiple Unit cars for the new Portland service. The diagrams on pages 106 and 108 illustrate the single level cars with the streamlined cab as well as the squared-end car. *From NEW DMU, Colorado Railcar, 2003 Edition, Page. 7*

Colorado Railcar is also building a bi-level for the Alaska Railroad.

Athearn has produced the Bombardier bi-level coaches and cab cars in HO gauge. These cars are operated on all but one of the Pacific Coast commuter railroad systems between San Diego and Vancouver. The only exception is Portland, which is purchasing equipment from Colorado Railcar. Athearn has also produced the F59PHI motive power, which is also operated by the West Coast Express, the Coaster,

Metrolink, and the Sounder. The models are accurate examples of the passenger equipment and motive power and provide modelers with a new level of the artwork of model railroading. As mentioned previously in this book, equipment from some of the commuter lines have been loaned out to other new services. Model railroaders have the opportunity to introduce new commuter services on their layout by using models from any of the routes produced by the model producers. For more information on modeling commuter train services, one may wish to check the December 2007 issue of *Model Railroader*.

The photos on pages 109 and 110 by Dan Mackey illustrate the Athearn's HO gauge model of the West Coast Express F59PHI, a Bombardier coach and cab car.

Single Level DMU Commuter Car
98 Seats
246 Maximum Passenger Capacity - Includes Standees*

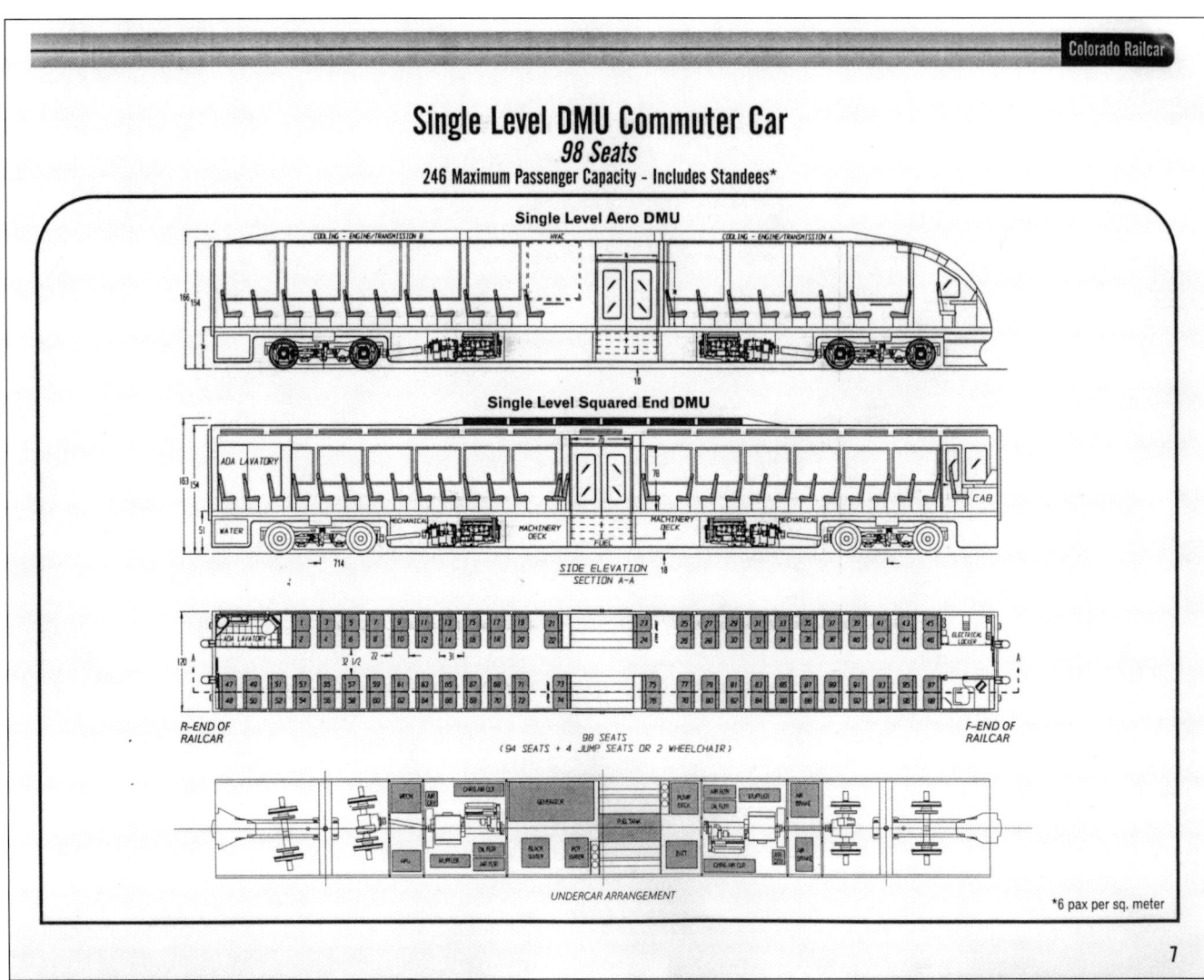

7

Colorado Railcar

The Colorado Railcar DMU is shown here on display at the Amtrak Midway Station at Minneapolis and St. Paul. *Patrick C. Dorin*

Single Level Low-Floor ADA Accessible Coach w/Cab
92 Seats
254 Maximum Passenger Capacity - Includes Standees*

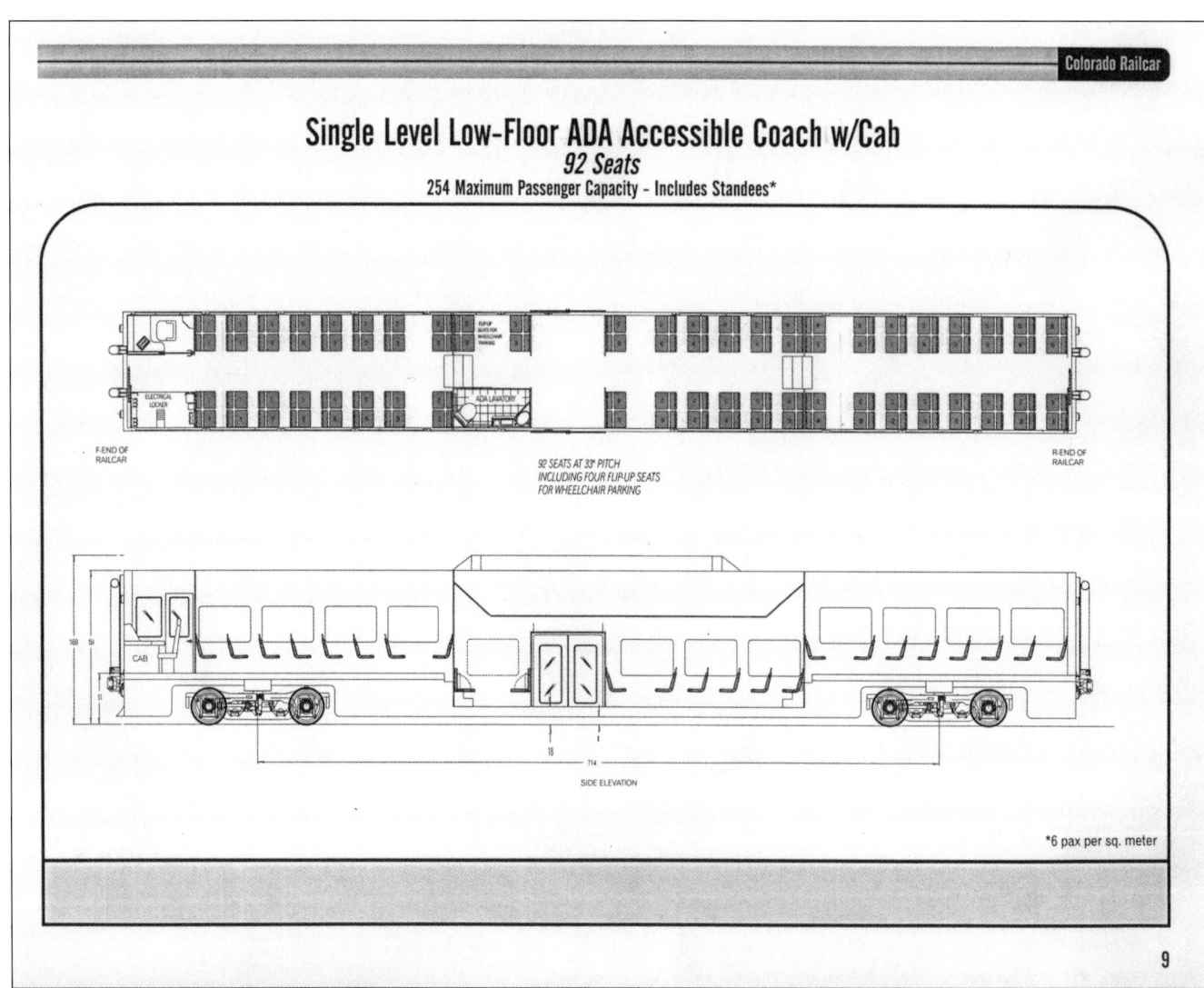

Colorado Railcar

A model two car with a coach and cab car pause at a suburban station. All model photographs by Dan Mackey.

Athearn's model F59PHI, the West Coast Express No. 903.

Model railroad commuter passenger equipment can provide city planners with a hands-on view of coaches and motive power, such as this roster photo of the West Coast Express model of car No. 212.

The West Coast Express cab car No. 102 by Athearn.

RAILWAYS

TRUCKS

BUSES

AUTOMOTIVE

More Great Titles From
Iconografix

All Iconografix books are available from direct mail specialty book dealers and bookstores worldwide, or can be ordered from the publisher. For book trade and distribution information or to add your name to our mailing list and receive a **FREE CATALOG** contact:

Iconografix, Inc.
PO Box 446, Dept BK
Hudson, WI, 54016

Telephone: (715) 381-9755,
(800) 289-3504 (USA),
Fax: (715) 381-9756
info@iconografixinc.com
www.iconografixinc.com

More great books from Iconografix

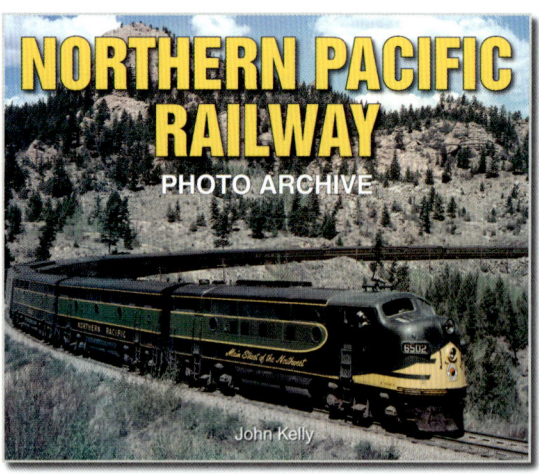
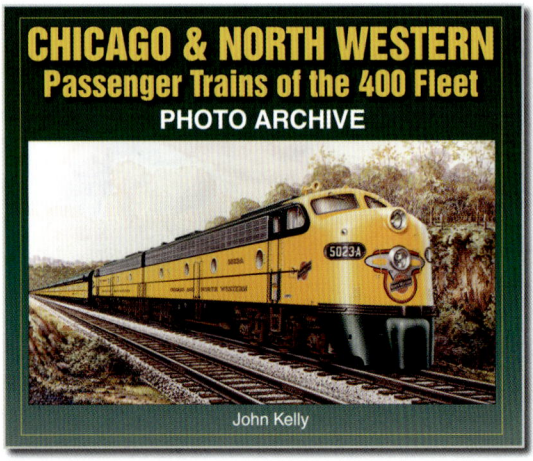

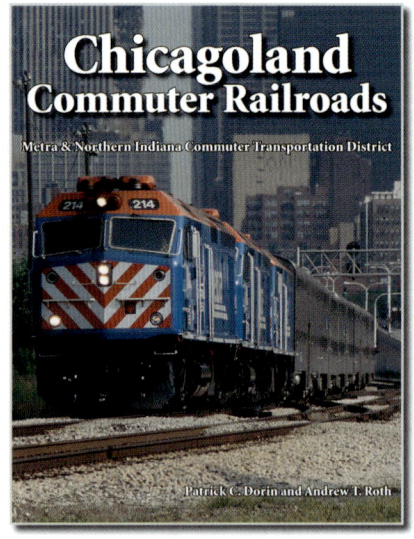